Cambridge Energy Studies

The realities of nuclear power

international economic and regulatory
experience

T0292190

The realities
of nuclear power

international economic
and regulatory experience

S. D. THOMAS

Senior Fellow, Energy Programme
Science Policy Research Unit, University of Sussex

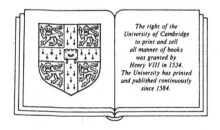

The right of the
University of Cambridge
to print and sell
all manner of books
was granted by
Henry VIII in 1534.
The University has printed
and published continuously
since 1584.

CAMBRIDGE UNIVERSITY PRESS

Cambridge
New York   New Rochelle   Melbourne   Sydney

CAMBRIDGE UNIVERSITY PRESS
Cambridge, New York, Melbourne, Madrid, Cape Town, Singapore,
São Paulo, Delhi, Dubai, Tokyo

Cambridge University Press
The Edinburgh Building, Cambridge CB2 8RU, UK

Published in the United States of America by Cambridge University Press, New York

www.cambridge.org
Information on this title: www.cambridge.org/9780521126038

© Cambridge University Press 1988

This publication is in copyright. Subject to statutory exception
and to the provisions of relevant collective licensing agreements,
no reproduction of any part may take place without the written
permission of Cambridge University Press.

First published 1988
This digitally printed version 2009

A catalogue record for this publication is available from the British Library

Library of Congress Cataloguing in Publication data
Thomas, S. D. (Steve D.)
The realities of nuclear power: international economic and
regulatory experience/S. D. Thomas.
    p. cm.
Bibliography: p.
Includes index.
ISBN 0 521 32750 4
1. Nuclear industry.   I. Title.
HD9698.A2T53 1988
338.4'7621483—dc 29   87–22488 CIP

ISBN 978-0-521-32750-3 Hardback
ISBN 978-0-521-12603-8 Paperback

Cambridge University Press has no responsibility for the persistence or
accuracy of URLs for external or third-party internet websites referred to in
this publication, and does not guarantee that any content on such websites is,
or will remain, accurate or appropriate.

Contents

Acknowledgements

I am greatly indebted to very many people for their contributions to this book. Firstly, I am grateful to those people who gave generously of their time and opinions in interviews – they number more than one hundred. Secondly, I am grateful to individuals who have commented on draft texts and who have contributed ideas. Here I should mention Clark Bullard, Charles Komanoff, Andrew Reynolds and Eli Roth who were particularly important sources of ideas. Thirdly, I would like to mention my colleagues in the Science Policy Research Unit Energy Programme particularly Gordon MacKerron, but also Sonja Boehmer-Christiansen, John Cheshire, Jim Skea, John Surrey and William Walker. Despite an endless flow of drafts demanding comments, they have continued to provide stimulating and challenging comments. Fourthly, I must thank relatives and friends for their continuing encouragement and interest. Finally I must thank Eunice Surtees-Hornby whose patience and typing skills made the final report possible, and Jane Whittingham who compiled the index.

The book has been made possible by a grant from the Science Research Council and subsequent support from the Economic and Social Research Council.

CHAPTER 1

Introduction

Nuclear power is now at a crossroads. New orders for nuclear plant have almost completely dried up and many of the technological, economic, political and energy policy premises on which nuclear power's attractions were based have proved unfounded. The expectations that factors such as learning and scale economies would mean that the cost of nuclear-generated electricity would become significantly and substantially cheaper through time have not been fulfilled. The performance of reactors remains disappointing and capital costs have increased steadily in real terms so that financing new reactors has become a severe, and in some cases, insuperable barrier to further orders.

In energy policy terms, mainly for long-term structural reasons, electricity demand has not grown as anticipated and the need to substitute for fossil-fuels, which seemed imperative after the first oil crisis, no longer seems so urgent. Politically, the scale of public opposition that has occurred was not anticipated and the expectations that fears would recede as they were either proved groundless or the dangers at least became familiar have not been fulfilled. The accidents at Chernobyl and Three Mile Island have ensured that any developments or incidents relating to nuclear power are extensively reported and that proposals to build nuclear power plant anywhere are likely to be subject to the closest scrutiny.

However, the hiatus in ordering would appear to give the opportunity to re-evaluate nuclear power thoroughly. Such a re-evaluation would be timely as a sufficient body of experience is now accumulating to allow nuclear power to be properly evaluated on the basis of actual rather than anticipated performance. From this experience, it should be possible to address a number of issues concerned with the performance of the technology. These would include:
– an assessment of the available technologies to examine whether they are the most appropriate and what improvements should be made to them

- a review, to establish how to ensure good practice in building and operating nuclear plant
- a re-evaluation of state-support for nuclear power, particularly the effectiveness of national atomic energy research organisations and national research budgets
- a review of the role and methods of the safety regulators
- an analysis of what adjustments are necessary to the nuclear power plant supply industry and what positive steps are needed to achieve them

Whether the past history of nuclear power has created so many rigidities and entrenched positions that the nuclear industry is incapable of the degree of change required remains to be seen but if past lessons can be thoroughly learnt it may be possible to ensure that the nuclear power industry is fully able to respond, if substantial orders for nuclear power plant are again required.

More generally the development of nuclear power technology raises and illustrates a number of issues on the development and commercialisation of large-scale technologies. It is especially relevant to technologies that are seen as being of strategic importance and which have such long lead-times and development costs that private industry has been unwilling to undertake their development alone. Some of the issues raised include:

- the importance of learning by experience as a means of technology improvement
- the impact of scale economies on costs
- the role of technology standardisation
- the incorporation and use of prototype and pre-commercial experience

The scope of the book

There is little understanding why, even though nuclear reactors have been supplying electricity to consumers for 30 years, there is still a wide variation in the success of nuclear power programmes. Some countries have been able to build nuclear plant quickly and cheaply, and operate it effectively, whilst others have been much less successful. This book examines the contrasting records of four of the major users of nuclear power, the USA, the Federal Republic of Germany, Canada and France, and seeks to identify the factors which have been important in determining the success or otherwise of their nuclear power programmes.

In particular, the operating performance record is examined highlighting the major technological weaknesses. Capital costs and construction times are also examined identifying trends and the underlying factors behind them. These technological and economic factors are then related to the specific environment for nuclear power in each country. This includes the constitution and strengths of the main organisations involved, such as the electric utility, the plant vendor and the safety regulator, the extent and form of government support and the form and rigour of regulatory activity prior to, and during, construction and during operation.

The world context

There are now about thirty countries worldwide operating or building nuclear plant and, for nearly all of them, this represents the most challenging project, technologically, logistically and politically, that these countries have undertaken. In technological terms, nuclear power requires the highest standards of design, engineering and operating practice and has imposed unprecedented regulatory requirements. Logistically, the enormous complexity of these plants creates problems in communication and co-ordination in the construction and operation phases.

The public image of nuclear power has changed radically over its history. In its early days, up to about 1970, public reaction to nuclear power was generally very favourable. The reasons for this are complex and difficult to summarise, but some of the main ones include:
- civil nuclear energy was a more acceptable fruit of the vast amount of effort poured into nuclear research than nuclear weaponry – in some way it salved the public's conscience over Hiroshima
- it appeared to epitomise modern technology producing the modern fuel, electricity, cheaply and also more cleanly than the fossil-fuel alternatives

However, public opinion had begun to polarise by the early 1970s. To its strongest opponents, nuclear power had not broken free from its military links (see for example Patterson, 1984) and, increasingly importantly, the safety hazards did not compensate for the benefits of apparently cheap power. By contrast, its strongest supporters remained unshaken in their belief that nuclear power was the only acceptable way of securing energy supplies for the long-term. Increasing costs and disappointing performance were ascribed to the action of intervenors and poor regulation, rather than to failings within the industry. This degree of polarisation, which has remained through the past 15 years, is a feature unique to nuclear power. For no other civil technology is it possible to categorise such a high proportion of the population as being 'anti' as a matter of principle.

Whilst it is quite legitimate to argue for or against a technology on grounds of principle, such arguments have dominated nuclear power and the quality of economic and technological argument has generally been low. Much of the analysis that has been produced uses the data to reinforce a prior position whilst more neutral analysis is branded as 'pro' or 'anti' according to the conclusions reached.

Two of the major factors to consider in evaluating nuclear power programmes are the variety and sometimes confusion of motives countries have had in adopting nuclear power and the route to nuclear power these countries have taken. Particularly in the early phases, military considerations were often of importance either directly by determining technology choice or indirectly by giving particular technologies a head start. (The importance of the links between the military and civil programmes is strongly argued by Roth (1982).)

In the USA, the light water reactors (LWRs), which now dominate the world market, were given a considerable lead over their competitors by their prior use as submarine propulsion units and the pre-existence of uranium enrichment facilities required by the weapons production programme. In the UK, the first reactors in the gas-cooled reactor programme were controlled by the government and were primarily designed to produce plutonium for weapons (see Williams, 1980). Electricity was a by-product.

In Canada, the civil nuclear programme has been least affected by military considerations, Canada having renounced nuclear weapons soon after the Second World War. However, even here, its programme of heavy water reactors was assisted by the experience it had gained in that war producing and supplying heavy water to US military establishments.

Nowadays in most countries civil and military nuclear programmes are largely separated, although suspicions remain about the motives of some of the countries developing nuclear power capabilities.

Nevertheless the result of this military legacy has been the continuance of large subsidies, often to government laboratories, to develop nuclear power technology further. This has meant that the ultimate users, the electric utilities, have often had little influence on the choice of technology and the direction of technology development. The early vision of the way in which nuclear power technology would develop has persisted even when its grounding tenets no longer seem valid. This vision held that 'thermal' reactors, that is, reactors which could only use a small part of their uranium fuel, would be rapidly succeeded by 'fast' reactors, which use a much higher proportion of the uranium (see Walker & Lonnroth, 1983). The abundance of uranium and the slower than expected development of nuclear power has meant that such a transition is now seen as increasingly distant if indeed it ever becomes necessary. Nevertheless, government funds still reflect past perceptions and are heavily concentrated on fast reactor development. Development of commercial reactor types has been carried out largely by private industry since the first commercial orders and has often followed a 'fire-fighting' reactive approach rather than an innovative approach.

As military imperatives have diminished, civil nuclear power programmes have been increasingly driven by energy policy and economic policy objectives. In the early to mid-1950s, before oil had acquired its status as a cheap, plentiful and reliable fuel source, nuclear power was seen in some countries as a natural replacement for coal as indigenous reserves were exhausted. This influence weakened through the 1960s as oil production from the Gulf States increased, returning to prominence following the first oil crisis of 1973–74 but since then waning yet again. A second energy policy objective, that of reducing the cost of electricity supply, was established by the milestone US Oyster Creek nuclear power plant order of 1963. Here, for the first time, an order was placed for a nuclear plant because, on the basis of bids received, it was the cheapest option available to the utility. (For further details see Bupp & Derian, 1981.) As with the 'fuel-shortage' objective, the attractiveness of this argument was weakened by falling real oil prices and rising real nuclear capital costs. It was given a new lease of life by the oil crisis before losing ground again as capital costs continued to escalate and oil and coal prices failed to rise as anticipated.

A third energy policy motive, that of increasing the diversity of electricity generating capacity, has emerged strongly since the second oil crisis. This has arisen because of fears that electricity supply systems that were heavily dependent on one particular fuel would be vulnerable if supplies of that fuel were interrupted or its price rose quickly.

Economic and industrial policy motives are often less overt but have nevertheless played an important role. The high-technology nature of nuclear power persuaded governments and companies that it was strategically important to establish capabilities in this field in order to pick up a share of what was seen as a rapidly expanding and lucrative world market and to keep abreast of what were seen as the skills and techniques of the future. More recently, rapid fluctuations in oil prices have increased the attractiveness of energy sources such as nuclear power which do not cause unpredictable strains on the nation's balance of payment. For many countries the cost of importing oil has been a major drain on resources and has been seen as a barrier to economic growth. However, the escalating cost of nuclear plant has diminished nuclear power's attractiveness in this respect. Countries such as Mexico, Brazil, Argentina, the Philippines and even France are finding that their nuclear power programmes are a major contributor to their international debt. (For example, a more detailed account of Mexico's debt problems can be found in *Nucleonics Week*, 1986c, p. 11.)

The strategies open to a country adopting nuclear power may be constrained by its underlying motive. For example, France decided to reduce dependence on imported oil radically and very quickly by substituting nuclear power for oil-fired electricity generation and, further, by substituting electricity for the direct use of oil. Given the scale of programme envisaged, probably the only viable strategy if costs were to be controlled and schedules met was to impose centralised, autocratic control, use standardised, apparently proven technology and build up a high level of indigenous component supply. In other countries, such as India and Pakistan, suspected connections between the civil and military nuclear programmes have restricted these countries' ability to purchase equipment and fuel on the world market.

Corresponding to these changes and differences in motives have been changes in the location of decision-making. This has lain variously with atomic energy (and weapons) research agencies, utilities, vendors, politicians and the public. These changes in location have often brought about a re-evaluation or change in the direction of nuclear power policy.

One further factor that has had an important influence on the way in which electric utilities view nuclear power has been the history of technical innovation in the electricity supply industry. Throughout this century there has been a steady and remarkable stream of innovations in the electrical power generation sector. Up to the 1960s thermal power stations were generally sited in city-centres and produced unpleasant local pollution. Economies of scale, improved pollution control equipment (especially for particulates), higher stream conditions at the turbine generators and improved transmission technology have all served not only to steadily reduce the real cost of electricity, but also to reduce the cost of electricity relative to other energy forms. In addition, it has allowed the increasing demand for electricity to be met by fewer power stations which have usually been sited in remote areas. By the late 1950s and early 1960s the 'all-electric' home had become an advertising cliché for advanced lifestyles. However, by this time generation technology was running up against diminishing returns. The benefits of building larger power stations, producing steam for the turbines at higher temperatures and pressures and transmitting electricity at higher voltages were no longer substantial. In this situation, to many in utilities and governments, nuclear power seemed to open up a whole new stream of innovation and challenges and, in a phrase that has rebounded on the electricity supply industry, promised 'power too cheap to meter'.*

On a range of other issues, electric utilities have failed to respond quickly to changing circumstances. Such issues include:
- the rising real cost of nuclear power. Utilities have tended to blame rising costs on disruption caused by intervenors, and unnecessary safety

---

* This phrase was used by Lewis Strauss, the then-Chairman of the US Atomic Energy Commission at a National Association of Science Writers' Founders' Day Dinner in New York (September 16 1954).

requirements imposed by hostile regulators. As is argued in this book, these may have been factors but they by no means account for all the real cost increases
- the long-term reduction in electricity demand growth rates. Utilities have tended to attribute these reductions to short-term interruptions caused by the oil crisis rather than long-term effects based on factors such as demand saturation and industrial restructuring away from heavy energy-intensive industries
- the scope for energy conservation. Utilities have been slow to realise the scope for energy (and electricity) conservation that energy price rises have produced. For many utilities, particularly where the costs of new plant are highest, the sponsoring of conservation schemes is a more cost-effective route to the utility than allowing demand to rise and building new plant
- the attractiveness of technologies competing with nuclear power. Such alternatives include fossil-fuel based technologies such as combined heat and power and combined cycle plant, and renewable technologies such as wind, tidal and wave power. Either explicitly or implicitly, more stringent financial targets were placed on these technologies than were applied to nuclear power and they have tended to be developed more slowly than they would merit
- the availability and attractiveness of fossil-fuels. Following the oil shocks, utilities have tended to overestimate how quickly direct use of fossil-fuel would decline in response to depletion and the supposed greater user-attractiveness of electricity. This has led them to overestimate the future price of fossil-fuels and underestimate their availability

These factors should have led electric utilities to adopt a cautious approach with respect to plant ordering, spreading research over a range of non-nuclear technologies and investigating demand-side measures such as conservation and load management.* However, many utilities have been reluctant to reduce their plant ordering. In the USA, because of the system of economic regulation of utilities which does not allow utilities to recover all the costs of unnecessary plant from consumers, this has resulted in a vast number of cancellations of nuclear (and other) plant. In countries with less stringent economic regulation of utilities, nuclear plant has been completed and has either not been fully utilised or has caused other plant to be underutilised or prematurely retired.

---

* Load management covers a number of options, but of most interest in this context are those which reduce peak demands on electricity systems by switching off (with the agreement of the consumer) non-essential uses. This can effectively reduce the amount of plant required to meet the same overall demand.

8   The realities of nuclear power

The current status of nuclear power

The current status of nuclear power is summarised in Table 1.1. Nuclear power contributes about 10% of the World Outside Communist Areas' (WOCA) electricity generating capacity and this proportion is likely to rise somewhat as nuclear capacity increases by a little less than 50% by 1995.*

Nuclear power in Communist countries. As was amply demonstrated by the Chernobyl accident, nuclear power programmes in the Communist world can have a profound effect on Western countries. However, little detailed analysis of the performance of these nuclear power programmes is possible. This is because few analytical data are available from which a worthwhile assessment of performance and procedures could be made. No cost data have been published and COMECON countries have not submitted the detailed accounts of performance to the IAEA on which the analyses of operating performance contained in this book are based.

Nevertheless, a number of points should be made about these programmes. Not surprisingly the dominant influence is the Soviet Union which has by far the largest number and capacity of reactors and which has supplied most of the plant in operation or under construction (see Table 1.1).

The Soviet Union's programme is based on two technologies, their own design of pressurised water reactor (PWR) and a light water cooled, graphite moderated reactor (LWGR) known as RBMK design (the design installed at Chernobyl). This latter design represents about 60% of installed nuclear capacity but only represents about 25% of capacity under construction. There are two current designs of RBMK type reactors, the RBMK 1000 and the RBMK 1500 (with capacities of about 1000 MW and 1500 MW respectively). However, no reactors of this design have been exported.

The PWR, which is likely to dominate future orders also comes in two sizes, the VVER 440 and the VVER 1000 (with electrical outputs of about 440 MW and 1000 MW respectively). All the exported units, including the only units exported outside the Communist Bloc, the two units at Loviisa in Finland† are of the VVER 440 design. For future orders the Soviet Union has recently withdrawn the VVER 440 design and all the PWRs under construction in the Soviet Union are VVER 1000s.

The only COMECON country to order reactors from a Western vendor is Romania which concluded a deal with AECL of Canada in 1978 for up to

* The proportion will rise because nuclear power represents a large proportion of plant under construction. Note also that some of the nuclear plant currently in service will be retired before 1995 although this is likely to be only small old plant and that some of the plant under construction may be cancelled.

† The operating record of the two Loviisa units is outstandingly good.

Table 1.1. *World capacity^a of nuclear plant*

Country	Main technology^b	Capacity (no. of units) Capacity (MW) at 1.1.86	Probable additions^c (MW) to 1.1.95
World Outside Communist Areas (WOCA)			
USA	LWR	77 106 (91)	29 893 (27)
France	PWR	33 365 (39)	27 581 (22)
Japan	LWR	23 640 (33)	8 736 (8)
FR Germany	LWR	16 068 (16)	6 846 (7)
Canada	PHWR (CANDU)	9 487 (15)	5 630 (7)
Sweden	LWR	8 325 (12)	—
UK	GCR	5 825 (22)	6 180 (10)
Belgium	PWR	5 465 (7)	—
Spain	LWR	5 577 (8)	1 902 (2)
Taiwan	LWR	4 984 (6)	—
South Korea	PWR	3 580 (5)	3 686 (4)
Switzerland	LWR	2 882 (5)	925 (1)
Finland	LWR	2 310 (4)	—
South Africa	PWR	1 844 (2)	—
Italy	LWR	1 285 (2)	1 964 (2)
India	PHWR (CANDU)	1 034 (5)	1 100 (5)
Argentina	PHWR	935 (2)	692 (1)
Yugoslavia	PWR	632 (1)	—
Brazil	PWR	626 (1)	1 229 (1)
Netherlands	PWR	481 (1)	—
Pakistan	PHWR (CANDU)	125 (1)	—
Mexico	BWR	—	1 308 (2)
Philippines	PWR	—	620 (1)
Total WOCA		205 576 (278)	98 292 (100)
Central Planned Economies (CPE)			
USSR	PWR, LWGR	25 977 (42)	34 725 (35)
DR Germany	PWR	1 702 (5)	2 448 (6)
Bulgaria	PWR	1 620 (4)	1 906 (2)
Czechoslovakia	PWR	1 570 (4)	4 068 (9)
Hungary	PWR	1 224 (3)	408 (1)
Romania	PHWR (CANDU)		1 887 (3)
Poland	PWR		880 (2)
Cuba	PWR		816 (2)
China	PWR		288 (1)
Total CPE		32 093 (58)	47 426 (61)
Total world		237 669 (336)	145 718 (161)

^{a} Capacity figures are expressed net of power station own use.
^{b} Main technologies:
    LWR = light water reactors including both pressurised water reactors (PWRs) and boiling water reactors (BWRs)
    PHWR = pressured heavy water reactors
    CANDU = Canadian deuterium uranium reactors
    GCR = gas (carbon dioxide) cooled reactors
    LWGR = light water cooled, graphite moderated reactors
^{c} Includes only plant under construction or firmly ordered; any orders not already placed are highly unlikely to have entered service before 1995. For plant in Communist countries, includes only plant reported as being under construction.
Source: Nuclear Engineering International, Power Reactors 1985, August 1985.

six units of which three are under construction. Progress on these units has been very slow and the completion dates consistently revised back.

China has been seen by many of the Western vendors as a major new market with expectations of 12 units, each of about 1000 MW being installed this century. However, contracts have been slow to be finalised and the only unit under construction is a 300 MW PWR of Chinese design. There is reported to be a commitment to purchase two 900 MW PWRs from Framatome of France but a firm order had not been placed at the time of writing. China now appears to be placing more emphasis on coal for future power generation than nuclear power.

Overall, the commitment to nuclear power in Communist countries seems stronger than in Western countries with nuclear capacity expected to increase by nearly 150% over the next ten years. However, this impression may be misleading. It reflects that relatively few reactors have been installed in Communist countries. In addition, these countries are subject to the same pressures, particularly financial, that have drastically reduced expectations of the amount of additional capacity that will be installed in the West. The difficulties of gathering reliable data mean, however, that the nuclear programmes of Communist countries are not considered further in this book.

The assessment of nuclear power

Despite the long history of nuclear power plant operation there has been little feedback of actual operating experience into decisions about technology choice. There are three main reasons for this.

First, long construction times and rapid product change have meant that, at least until the mid-1970s, the reactors available for order bore little resemblance to those in operation. This meant that it was hard to draw any meaningful conclusions, based on actual experience, about the merits of various reactor systems until the mid-1970s. By this time most countries that are now users of nuclear power had already chosen their technological route. The choice of reactor system often led to the construction of dedicated production facilities, the training of high-grade engineering and scientific talent to operate with the given design and the expenditure of considerable funds to research the system and set up a well-trained regulatory establishment. Such decisions and actions tend to generate a momentum which is difficult to halt or reverse.

Second, due to safety implications and the demanding nature of the technology, some countries have been wary of adopting technologies other than the market leaders. These seem to give greater security and back-up in the event of something going wrong, even where the alternatives may appear more attractive.

Third, as was discussed earlier, the highly controversial nature of the technology has meant that the independent analyses which were produced were frequently not examined objectively. Industry, utility and government sources have often been biased and have tended not to come to conclusions which might reflect critically on them or be used to undermine the existing policy position.

For much of the commercial history of nuclear power, decisions on technology choice were taken on engineering or economic criteria which were not based on actual experience. For example, in 1964 in the UK, one of the reasons the electricity utility, the CEGB, chose to order advanced gas-cooled reactors (AGR) rather than boiling water reactors (BWR) was that the former's ability to be refuelled while it was operating appeared likely to give it better annual availability which counter-balanced its higher expected capital costs. This was expected to give the AGR a marginal advantage in total generation costs. Similarly, a large number of reactors of about 1200 MW capacity were ordered in the USA on the basis of expected economies of scale before any reactor of even half that size had been commissioned. That such theoretical advantages should be treated with utmost caution when making far-reaching decisions on technology choice has been amply demonstrated by the economically disastrous experience of the UK's AGR programme, and by the difficulty of detecting economies of scale in actual outturn costs in the USA.

By the mid 1970s, attention was beginning to be drawn by independent analysts to the economic performance of nuclear plant. These analyses, which covered units in the United States, showed that, in general, operating performance was falling well short of the levels expected and that capital costs were escalating at an alarming rate. In addition, nuclear plant was not following the sort of trajectory that theories such as those based on 'learning-by-doing' and technology maturation would have suggested. Other analysts took up this work expanding it to include reactors worldwide and also looking in greater depth at the factors causing poor performance.* These analyses confirmed that overall performance was disappointing. They revealed consistent and substantial differences between the various technologies, between vendors and between different sizes of plant. Such differences were not expected but are reasonably easy to rationalise in terms of technological characteristics; put simply, product A is better than product B. However, these analyses showed a second set of differences,

* The pioneering analyses in this area were carried out by Komanoff (1976). Joskow and Rozanski (1979, pp. 161–8) attempted to relate plant performance to general economic theory. Surrey and Thomas (1980, pp. 3–17) examined units worldwide detailing the causes of plant unavailability. Since then numerous analysts have examined plant performance in various ways.

apparently similar units were performing far better in some countries than others and even in different utilities within the same country.

This second set of differences, which are as large as the technological ones, can only arise from factors specific to countries and individual reactors. Such factors would include the quality of the components, how well the plant was built, operating and maintenance of the plant, and the style and methods of the regulatory agencies. Initial explanations, for example that poor performance was attributable to hostile regulators and public opposition and that good performance required a large utility, or a country with strong traditions of engineering excellence, have proved naive. For example, if US regulation and public opposition is so disruptive to nuclear plant operation, how is it possible that some plants (often operated by small utilities) are able to produce performance, year-in year-out, which is amongst the best in the world? Similarly, if what is required is a country with a strong engineering industry, how is it that whilst in FR Germany, PWRs have performed outstandingly well, their programme of BWRs has experienced severe difficulties?

For any process of technology acquisition to succeed, however simple the technology, it is not only necessary to choose a sound design but also to build, operate and maintain it well. For nuclear power stations it is evident that the balance of difficulty lies very much more with the building, operation and maintenance aspects than, for example, is the case for coal-fired power stations.

Whilst it is certainly true that no-one in the nuclear industry thought, at the outset, that building and operating nuclear plant was likely to be easy, few can have anticipated the complexity of problems which have confronted the industry. This has meant that those involved in nuclear power, notably plant vendors, utilities, regulators and government, have had to respond continually to new issues and challenges, in particular those of establishing a responsive regulatory system, controlling escalating construction times and costs and improving often disappointing operating performance.

The case study countries

More than 85% of capacity in operation or under construction in the non-Communist world is sited in the seven largest nuclear power using countries: the USA, France, Japan, FR Germany, Canada, Sweden and the UK. The four countries discussed in this book, the USA, Canada, France and FR Germany, epitomise the challenges and problems that have faced nuclear power, each illustrating a different aspect of the problems. The USA, the source of the market-leading technologies, has experienced rapidly escalating construction costs and times, highly variable operating performance and, at times, severe public opposition. In contrast, Canada which has taken an independent technology route, has until recently achieved consistently outstanding operating performance, relatively stable construction costs and times and has experienced little public opposition. However, the extent to which this achievement is due to the uniquely close institutional arrangements between utility, vendor and regulator and how much it is due to the intrinsic quality of the basic technology is not easy to establish.

FR Germany has adopted the same basic technology as the USA but with some important modifications. As in the USA there has also been severe public opposition which has had an adverse effect on construction times and costs. However, at least in part, operating performance has been very good despite an irregular flow of orders to the vendor and the absence of 'production line' facilities for reactor components.

France, which was one of the pioneers of nuclear power with its own technology, adopted American technology much later than Germany, setting in motion an unprecedentedly large programme of standardised reactors. This has created considerable interest because of its scale and because it represents the first major test of the merits of standardisation. Standardisation has been seen worldwide as a major force in containing construction cost and times, shortening licensing procedures and simplifying repairs and maintenance.

Of the other three countries, the UK is of limited relevance because its reactors were nearly all ordered 20 years ago or more using technologies, namely Magnox and its successor AGR, that are highly unlikely to be adopted elsewhere. Sweden is of more interest technologically having developed its own design of BWR independently. However, following a referendum decision in 1979, all reactors are expected to be phased out by 2010 and no more reactor orders will be placed. Japan is of considerable importance because of the scale of its programme and also because of its proven ability to import and develop technologies. However, published information and in-depth analysis of the factors underlying Japanese decisions are difficult to obtain and therefore Japan is not examined.

CHAPTER 2

A framework for evaluating nuclear power technology development

Introduction

This chapter sets up a framework in which the development of nuclear power technology can be analysed drawing upon general theories of technology development and illustrating the special features of nuclear power. Four processes or sets of processes which can be considered separately are identified; (i) the development of the technology including the utilisation of testing, research and the experience of producing or using the technology; (ii) the process of transferring the knowledge of how to produce and operate the technology to another producer; (iii) the development and constitution of the companies involved in producing the technology including their ownership, the scope of their operations and production methods; and (iv) the process of economic and technical improvement including technological change. Two further issues, standardisation and economies of scale, are dealt with separately because they have been particularly powerful influences on the direction and speed of nuclear power development.

Prior to setting up this general framework, it is useful to define the tasks involved in the various stages of a nuclear power plant's product life-cycle.

Designing, building and operating a nuclear power plant

The operations involved over the entire project life of a nuclear power plant can be split into nine activities:
- R&D into the technology to establish its viability or to build understanding of the logic of the design if the technology was not originated in the country involved
- design and manufacture (or procurement) of the nuclear steam supply system (NSSS), carried out by the vendor
- design and procurement of the remaining plant and its integration with the NSSS into the whole station design, performed by the architect engineer

- building the plant, a task supervised by the constructor
- operation and routine maintenance of the plant, the responsibility of the utility
- nuclear fuel supply and waste fuel management
- reactor servicing, particularly non-routine maintenance and repair
- ensuring the safe operation of the plant, the responsibility of the utility operating under the guidance and supervision of the safety regulatory authorities
- reactor decommissioning

The boundaries between these activities are not always clear-cut, particularly that between routine maintenance and servicing. Nevertheless, the main areas are well-defined and represent a useful sub-division.

These direct activities are heavily influenced by the political environment. Political influences may be explicit through specific decisions and strategies and through the operation of the safety regulatory agency (invariably responsible to the state). Two particularly important factors are the extent of state support, and the speed and certainty of procedures in selecting, and obtaining consents for, sites. Political influence may also be felt less explicitly, but equally importantly, through the political culture of the country. This may reflect policies on attitudes to factors such as centralisation, state-control and militarism.

Returning to the direct activities, a major feature distinguishing the four countries studied in this book is the way in which these tasks are allocated and performed. Table 2.1 summarises the institutional structures of the four countries with respect to five of the activities identified. The detailed structure of the institutions and the full significance of some of the terms is explained in the chapters on the individual countries but a number of examples will illustrate this diversity of approach.

In France, the vendor, Framatome, has set up comprehensive in-house manufacturing facilities, with most components being produced in its own factories using sophisticated production-line technology. The American vendors had followed this strategy in the 1970s but the dearth of orders in the past decade has led to the closing of many of these facilities and greater reliance on sub-contracting. The German and Canadian vendors have consistently followed a policy of sub-contracting manufacture of all the major nuclear components.

In the field of architect-engineering, the contrast is equally strong. In the USA, specialist companies are used generally, whereas in France and Canada, the utilities carry out their own architect-engineering and in FR Germany the vendor performs this function.

With safety regulation, the differences are often in style. In the USA, as with most fields of public policy, the style is very open and disagreements between utilities and regulators are publicly aired. In FR Germany there is

Table 2.1. *The institutional structure of nuclear power*

	USA	FR Germany	France	Canada
Vendors				
Number	3/4	1/2	1	1
Origins	Long-standing, diversified	Long-standing, diversified	Specially created	Specially created
Source of technology	Indigenous	Imported	Imported	Indigenous
Manufacturing capability	Increasingly sub-contracted	Almost entirely sub-contracted	Almost complete in-house capability	Almost entirely sub-contracted
State support	None direct	None direct	Underwritten by the government	Owned by the government
Architect-engineers	Specialist companies (occasionally utilities)	Vendor	Utility	Utility
Constructors	Specialist companies (occasionally utilities)	Specialist companies	Specialist companies	Utility
Utilities				
Number involved	Over 50	Approximately 10	1	3 (one dominant)
Size	Wide range	Medium/large	Very large	Large
Ownership	Generally investor-owned	Investor-owned	Publicly-owned	Publicly-owned
Catchment area	Variable	Generally regional	National	Provincial
Safety regulation				
Utility relations	Arm's length	Arm's length	Close, co-operative	Close, co-operative
Public disclosure	Very extensive	Extensive	Limited	Extensive
Methods	Active	Active	Reactive	Reactive
Centralisation	Partially devolved	Devolved, pluralistic	Centralised	Centralised

also fairly full disclosure but with the added complication that much of the responsibility for safety regulation is devolved to the individual *Land*. (FR Germany is a federal state of 11 *Länder*.) In both countries, the regulators tend to be more active (as opposed to reactive) in the design phase specifying features of the design. In Canada and France relations between safety regulators and utility are very much closer, with negotiations on safety carried out behind closed doors, and the resulting decisions are presented to the public behind a combined front. The regulators also react to designs presented to them suggesting modifications rather than pre-specifying design features.

There are four activities defined previously but not covered in Table 2.1. The prior R&D, which is discussed in detail in the case study chapters, is a complex process with responsibilities changing through time and cannot be summarised readily. The remaining three activities – the nuclear fuel cycle, reactor servicing and reactor decommissioning – are not central to the themes of this book for a variety of reasons. Variability in the quality of fuel cycle services does not seem to have been a major factor in determining reactors' economic performance. However, failure to make adequate provision for the back-end of the fuel cycle has occasionally affected planning and the operation of individual plants, and, where this is the case, it is noted.

Reactor servicing is of more direct relevance to reactor economics, but the market is still at an early stage of development. In this context reactor servicing is defined to include non-routine maintenance and repairs. As more reactors reach 'middle-age' and 'old-age' and original equipment wears out, this market will assume increased prominence. Evidence of the importance of this field can be seen, for example, in the growing competition for work in servicing and replacing steam generators and in decontaminating plant. Because of the diversity of the tasks involved, it is unlikely that a pattern of institutional involvement will emerge as clearly as for the other activities. The evidence to date is that servicing will be a very competitive field. Utilities are seeking to increase their revenues, capitalising on their own experience by selling their services to other utilities. Plant vendors are seeking to maintain a flow of work to their currently under-utilised design and research resources and also seek to establish or maintain strong links with utilities by offering reactor services (not just to reactors of their own make). Specialist companies see an opportunity to develop a market niche with highly specialised skills which can be applied world-wide.

Reactor decommissioning, although increasingly the subject of public concern, has not been firmly established in terms of techniques, timing and responsibility. It will not be until the first crop of commercially-sized reactors (larger than 500 MW) become due for retirement early in the next century, that the economic and environmental impact of decommissioning can be judged.

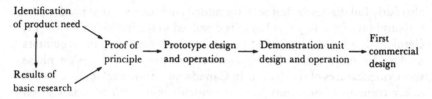

Figure 2.1. The early development of a technology.

THE FRAMEWORK FOR ANALYSIS

Although there is an extensive literature on technological progress and development, a common problem with many of the terms used is the lack of widely-agreed precise definitions for them. Terms such as 'learning' and 'technological change' are frequently used with varying degrees of precision and meaning. This is not necessarily of particular importance provided the terms are used consistently and in a clearly defined manner. Where such ideas are introduced in this chapter, a working definition of how they will be used in this book is given.

The development of technology

A highly simplified representation of the development of a technology in five stages up to the production of the first commercial* design is shown in Figure 2.1. The first stage, identification of product need/research results is the most variable phase. The impetus can either be provided by the identification of a product need, by a user or a producer,† leading to research to produce such a product, or the results of research might suggest a useful product. The next stage is generally proving that such a product is feasible on a laboratory scale. From this experience a small-scale prototype can be built and used to identify weaknesses and the potential for design improvement. The fourth stage involves the design, construction and operation of a demonstration unit at or close to commercial scale. Finally, the first commercial design can be produced and in this, the major technological problems should have been solved. At all stages, the product should be subject to increasingly stringent testing against cost criteria and competing products, as basic cost parameters can be more accurately

---

* Commercial is defined as being proven and self-sustaining, requiring no subsidies or state interventions.

† Producers are here defined as anyone involved in the design and manufacture of the product whilst users are those who purchase and use the product.

Design A   Design B

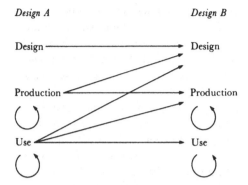

Figure 2.2. The incorporation of learning. The arrows represent the feedback of experience.

predicted. This process of evaluating and other factors, such as the development of more attractive competing technologies and the emergence of unexpected technological difficulties, mean that there is a considerable chance that the technology could fail at any stage.

Whilst examining this phase of nuclear power development is not a prime focus of this book, two important factors will be noted:
- was the source of the early impetus which led to the prototype stage user or producer led?
- how thoroughly were the lessons of the prototype and demonstration phase learnt and was there short-circuiting of any stages?

The subsequent stages of product development are shown, again in highly simplified form, in Figure 2.2. Here the life-cycle of a product is represented as three stages, design, production and use. In this diagram the role of 'learning' in improving and developing the product is stressed. Learning is here defined as any process whereby the experience of designing, producing or using is utilised to improve design, production or use of the product.*

The learning process is shown to operate from one design to the next, for example, the experience of using the first design should be fedback to improve the subsequent designs. Where subsequent designs differ in important respects from their predecessors, incorporating major or 'radical' changes, the subsequent designs are said to be of a new 'design generation'.†

---

* As originally defined by Arrow (1962, pp. 155–73), learning was a tightly defined, highly restricted process whereby the productivity of a production process was improved without capital investment. Since then, the use of the term has broadened and a variety of definitions has been used.

† It is important to note that a design may become radically changed by a number of incremental steps, none of which is, by itself, very far-reaching.

In addition, learning can operate in two ways within the same design generation. Firstly, the experience of using a given design may suggest small or 'incremental' changes to the production process or the design, which can be incorporated in later units of the same basic design. Secondly, the design of existing units may be capable of modification by replacing sub-systems during maintenance.

The key to the process of learning is the efficient communication of information and experience amongst users and producers. At its simplest the communications may be the reporting of experience, for example the occurrence of an equipment failure. At a more purposeful level, users might advise other users and producers of more effective methods of using the equipment. Information might also be prescriptive, for example the user specifying to the producer, using his experience, the specification for a new design.

This process of communication of experience is often clearly in the interests of users particularly in the case of electric utilities where they are not in direct competition with each other. However, for the producers of a technology, it will seldom be in their own narrow interest to share experience.

Two other processes of importance to the development of a technology are not shown in the diagram: testing and simulation, and research. These processes can have an influence at any of the stages of the product life-cycle.

Testing and simulation are methods increasingly used to bypass or short-circuit experience loops. The use of the product is modelled, either physically in which a scale or full-size model is exposed to the harshest conditions it will meet, or in a computer simulation in which the product is mathematically represented and tested for its response to certain conditions. Computer simulations are also of increasing value in operator training.

Before applying these ideas, particularly those on learning, to nuclear power, a number of modifications and qualifications must be made to take account of the special features of nuclear power. The factors which must be incorporated are:

- the important role of safety regulators and the public in influencing design
- the large number of institutions involved in each loop
- the long lifetime over which nuclear plants must be durable (perhaps 40 years or more) and the long lead-time between design and first operation
- the difficulties of product proving
- the disincentives to design changes

Safety regulation and public influence. For most products, the major impetus to technological change is to improve the economic or technical performance (or a combination of both). Safety requirements are a common factor for most products but they are often not a central feature of the design process – the product is designed and then made safe by the addition of extra features. For example, in the motor industry, many of the safety features now being incorporated, such as laminated windscreens and anti-lock braking, can be readily added at a late stage in the design process and the main force behind their incorporation has been used to increase the attractiveness of the product to potential buyers.

Whereas with many goods, particularly consumer goods but also some capital goods such as aircraft, the public's concern is mainly towards the safety of the operators/users, with many large capital goods such as nuclear power plant and chemical plant, the public's concern is towards the potential effect on the general population. In the former case where the public has some choice over the product and can make trade-offs between cost and safety, it is clearly in the interests of the producer to match its product to the perceptions of the public. In the latter case, the public has limited influence on the technology chosen and will, inevitably to some extent, have the trade-off between safety and cost imposed on them.

The unique risks posed by nuclear power and the possible interaction with military applications has created an unprecedentedly difficult set of issues for the public, and their political representatives at all levels, to handle. This has resulted in a continuing debate on the separation of civil and military aspects and on safety standards.

The large number of institutions. With many products there are basically only two sets of actors involved, the producer who designs and manufactures the product and the user who uses the product. For nuclear power, as was noted above, there is considerably greater variety in design influences. In addition, as is reflected in Table 2.1, there is a much larger number of institutions/companies at the production stage. For example, in the USA, utilities, vendors, architect-engineers and constructors are all involved in the production of the nuclear power station. The problems of ensuring a good flow of information between all these actors, bearing in mind the constraints of commercial confidentiality are formidable. Even in cases where no such constraints apply on the exchange of information, such as between users, it is only in recent years that the importance of feedback has been fully understood. This is illustrated by the recently growing influence in the USA of the Electric Power Research Institute (EPRI) and the Institute of Nuclear Power Operations (INPO). These are utility funded institutes which have as a major part of their function the efficient collection and exchange of experience.

The long product life-cycle. The long lead-times associated with nuclear power plants mean that it may take ten years or more before a major design or materials change can be tested in an operating reactor and many more years before it can be confidently assumed that the new design or material is durable and effective. This places a heavy burden on testing and simulation and means that learning-by-experience of design will be slow. However, commercial pressures to improve the product will be strong and there will be pressure to make design improvements, utilising the considerable volume of research that is being carried out, before previous design changes have been proven. This process of technological 'leap-frogging' may mean that aspects of a design may be obsolete before it enters service.

These difficulties of completing the feedback loop were particularly acute in the USA where reactors of 1200 MW were ordered in 1965–67 well before any operating experience had been gained with units even half that size. Only the broad shape of the design was fixed at the time of order and design changes, necessary either to incorporate new operating data or new safety requirements, were made throughout the construction phase which in some cases is only now being completed. Thus, the plant entering service now is a hybrid, incorporating safety and technological design changes generated over a twenty-year period.

The difficulties of product proving. A number of factors make the design of a nuclear power plant uniquely difficult to prove. These include the long lifetime (already referred to), the high product cost, the small number of units operating and the safety implications of product failure. The high cost of units and the safety implications of product failure mean that destructive testing is not possible for the whole system. Only the destructive testing of sub-systems and computer simulations of the whole system are possible. The high degree of interdependence of sub-systems in a nuclear power plant means the former type of testing may not show up the full consequences of a single failure, whilst computer models may not be a sufficiently accurate representation of the plant to anticipate all problems.

Radioactive contamination makes nuclear power plant uniquely difficult to repair and maintain, with many areas of the plant effectively barred to human entry. This has resulted in a great deal of development in robotic repair techniques and sophisticated remote methods of fault diagnosis. Despite this, some areas remain effectively impossible to inspect and certain equipment failures are regarded as irreparable. In addition, modifications to the design once the unit has entered service may be very restricted.

Figure 2.3. Technology acquisition.

The small number of units operating and the fact that there has been very little standardisation of plant makes it difficult to build up statistical analyses of product performance, which might help give a more rounded view of plant reliability.

The disincentives to design changes. This factor arises partly from the previous set of arguments, namely that design changes that are not successful will tend to be very expensive to remedy, and partly from the difficulty of obtaining regulatory approval for design changes. In countries such as the USA and FR Germany any design change will require lengthy and costly procedures to obtain regulatory approval, and may even open up areas of the design which had previously been approved to renewed, costly and perhaps unproductive scrutiny. Thus, even where designs are known to be faulty, it may be more economic, at least to the vendor if not ultimately to the user, to leave the design unchanged.

Overall, in a technology as complex as nuclear power, one of the most powerful potential forces for technological and economic improvement is the incorporation of experience at all stages in the product life-cycle, wherever it can be profitably used. However, the factors discussed above, such as the long product life-cycle and the large number of institutions, mean that the process of learning is very much more demanding than is suggested by the simple scheme shown in Figure 2.2. In addition, it must be stressed that many of the learning effects identified here are reversible. If capabilities acquired are not used there is a strong risk that this capability will be eroded and even lost.

The transfer of technology

The process of acquiring a technology is illustrated diagrammatically in Figure 2.3 showing three stages in ascending order of technological capability acquired, and commitment to the product.*

---

* A fourth stage, production of the product locally by a subsidiary of the originating company, exists but has not been of significance with nuclear power.

– *Direct import.* This is the lowest level of technology acquisition. The product is imported as a 'package' with the only substantial 'local' content coming, perhaps, from the assembly process. The importing country will acquire some knowledge of the technology, particularly its operation and maintenance, but will only acquire a limited capability in design and manufacture
– *Technology licence.* In this case a local vendor negotiates a licence for a specified period from the original vendor which allows the licensee, for a fee, to produce products identical or very similar to the original vendor's design. The terms of technology licences vary but, typically, give the licensee access to the licensor's new research and allow personnel to be trained by the licensor. The markets in which the licensee may operate and the extent of design changes allowed may be restricted under the terms of the licence
– *Independent capability.* In this case the vendor acquires a completely independent capability to design and produce the product

A possible strategy for acquiring an independent capability might involve progression through all three stages. A number of products would be directly imported and their performance assessed in operation to determine whether the product was worth acquiring and which of the competitors was best. Following this, by some process of selection such as market forces or government policy, a smaller number of designs would be licensed by local companies (perhaps those with prior links to the potential licensors) and further evaluated. Further selection would ultimately lead to some of the local companies breaking free from their licences (or not renewing them) and producing products distinct from those of the licensor.

The capability acquired by an importing country can be subdivided into two components, manufacturing capability and design capability. These can be seen as largely independent, that is, it is quite possible to imagine a country acquiring the capability to design a product and not the capability to manufacture it and vice versa. Generally the costs of building a manufacturing capability are direct, requiring factories, production lines and training for operatives, whilst the costs of building a design capability are primarily opportunity costs, requiring the recruitment and training of talented design staff who otherwise would be engaged on other challenging design projects.

A manufacturing capability is more difficult to maintain if a flow of work is not available to sustain it. It will typically involve a larger workforce than a design team and use facilities that will fall into disrepair unless operated. A design team will be cheaper to maintain if it cannot be fully utilised although it will also tend to break up unless meaningful tasks are found for it.

In practice, this idealised scheme of technology transfer may not be followed and the process may be modified by factors such as the level of demand for the product, an assessment of the potential and status of the product, the skills available in the importing country, and the speed with which it is desired to acquire a full capability.

For some countries, it may be appropriate not to proceed beyond the stage of directly importing the products. Such a decision would be taken by countries which:

– foresee a limited demand for the product, insufficient to guarantee full use of production facilities
– do not wish, or are not capable of committing the resources, either financial or skills, to take the process of acquisition to a higher level
– are not convinced of the long-term requirement for the product

Even to countries which do not meet these criteria, a policy of limited acquisition has attractions in terms of flexibility, low risk and low investment costs. In particular these include:

– ability to compare a number of competing products, buying whichever is most attractive financially and technologically
– low set-up costs
– greater ordering flexibility since manufacturing facilities needing a steady loading do not exist
– access to the worldwide state-of-the-art technology
– ability to purchase standard components on a competitive world market

The disadvantages are primarily in lost opportunities and lack of influence, in particular:

– inability to adapt the technology to the country's special needs and inability to make design improvements
– less influence over prices paid
– adverse balance-of-payments consequences
– little development of local skills

The pros and cons of the highest level of indigenous capability are generally the converse of those stated above. In particular a high level of indigenous capability will:

– allow the country to adapt the technology to its own needs
– give greater control over the cost of equipment
– minimise import content
– allow the development of new high-technology industries

Against these advantages, the following disadvantages exist:

– lack of flexibility. Changing technology choice, if it does not meet expectations, or alternatives become more attractive, will be costly and politically difficult

- considerable financial and manpower resources will have to be invested to establish the capability
- orders may have to be placed when not strictly required in order to maintain a healthy technology supply capability

Technology transfer and nuclear power

These issues of technology transfer are of particular concern with respect to nuclear power because of its perceived strategic importance and because of the scale of resources required simply to reach the lowest level of technology acquisition of importing and operating a nuclear power station. The long lead-times involved in nuclear power mean that it might now take two to three decades to pass through every stage of technology acquisition. In addition, few countries have sufficient demand for nuclear power plant to allow competitive evaluations and to justify fully independent capability in reactor production. These factors mean that only France with its PWRs and FR Germany with its BWRs and PWRs have followed the scheme of technology acquisition shown in Figure 2.3. Other countries have chosen not to acquire full capability, or are still acquiring the technology.

Small advanced countries such as Switzerland and Finland with a very limited need for nuclear power stations appear to have chosen to acquire a capability to comment critically on designs proposed to them. This allows them to be well-informed buyers with few of the rigidities a manufacturing capability would have created. By contrast large developing countries such as South Korea and Brazil have chosen to maximise local content in manufacturing, setting up production facilities whilst still remaining dependent on overseas vendors for design work. Only India amongst the developing countries has adopted a policy of pursuing early technological independence. As is explained in Chapter 7, this policy was not entirely voluntary and appears to have left India with a technology which may have features which are attractive to it* but may be technologically obsolescent.

Of the larger developed countries, only Sweden has gone straight to stage 3 of the technology transfer process developing its own design of BWR independently without a technology licence. France, FR Germany and Japan chose to license technology but with different speeds of transfer. FR Germany opted for early independence in both PWRs and BWRs and its design of both are now distinct from those of their licensors. France, and

---

* Such features include the use of natural uranium which avoids the need to build enrichment facilities or be dependent on foreign suppliers, and the output size of about 200 MW which may be more appropriate to a country with a relatively weak electricity grid than the reactors of about 1000 MW which now dominate.

particularly, Japan adopted a more cautious strategy, maintaining long-term links with their licensors.

The vendor's viewpoint

This account has so far concentrated on the perceptions of the technology importer. However, a different set of perspectives are held by the vendor exporting complex technological equipment. It is clear that the value placed by a vendor on export sales goes far beyond the simple profit and loss balance of the sale. One of the factors justifying export sales of complex products is to smooth fluctuations in manufacturing schedules. In periods of low home market demand, manufacturing facilities are kept in operation by winning export sales. These sales may be made at marginal or sub-marginal cost but the vendor may feel that this is justified if it maintains a manufacturing capability which might otherwise be endangered.*

Particularly for long-life, complex technologies, initial low profits or even losses may be recouped in servicing the product and supplying spares, areas in which the user is to some extent captive to the original vendor. More subtle, long-term strategies for an exporter include the creation of a protected market in which more profitable sales can be made and the use of an initial prestige sale as a 'beachhead' from which other goods produced by the same company can be sold. The objectives here are to establish the presence of the vendor in the country, promoting close user–producer relationships. This may mean that for future sales, the user will give preference to the same vendor partly because of good experience with the early sale and partly because of the knowledge of that vendor's products built up in the user. This non-price advantage should allow the vendor to charge higher, more profitable prices. In addition, if the vendor is a diversified company, it may use the prestige gained from a large project as the basis to increase sales into the same market of other goods.

Similarly, governments may underwrite or directly subsidise export sales to open up markets for goods from their country, or to gain some political advantage or influence in that country. Whilst these spheres of influence are sometimes established by successful technological competition, they also (perhaps more commonly) result from traditional, often colonial, links. For example, the UK has for a long time found former and existing Common-

---

* This factor appears to have been important in the market for conventional heavy electrical equipment such as turbine generators where it has been shown that the level of export activity of the major vendors was often related to the level of home demand. (See Surrey & Chesshire, 1972.)

wealth countries, such as India, South Africa, Australia and Canada, lucrative export markets.

A third set of objectives is much more related to prestige and technological leadership. Export sales, particularly to technologically sophisticated countries, may be seen as a powerful endorsement of the quality of the technology. Ultimately, as was noted previously, a user country which sees a long-term steady domestic demand for a technology is likely to want to build up a home capability for producing that product. The exporting vendor must accept this as inevitable but if it can substantially widen its experience base, benefiting from the new users' experience it can strengthen its technological position, perhaps at the expense of some manufacturing capability if the new user sets up its own manufacturing facilities (which may compete with it in world markets). The risk with this strategy is of setting up a powerful competitor in world markets which, having milked the exporter of its own knowledge, goes on to win orders in competition with it in world markets.

Technology sales and nuclear power

In relating these considerations to nuclear power, two special factors need to be borne in mind. Firstly, the volume of sales of nuclear plant is very low and, once a home vendor has been established, imports are unlikely as they would threaten the existence of the home vendor. Secondly, more than almost all other technologies, international trade in nuclear power equipment is politically constrained. Sales of equipment are scrutinised to guard against nuclear weapons proliferation and intergovernmental relations may well determine whether a sale is feasible.

Of these three motives for exporting – maintaining a steady flow of work, establishing protected markets and gaining technological advantage – the last two have been of most significance in the market for nuclear power equipment. The low number of orders placed and the increasingly fierce competition for orders has meant that vendors are in no position to pick and choose which orders they should compete for. This fierce competition even for single orders, however, does underline the importance vendors place on establishing protected markets. For example, Kraftwerk Union of FR Germany now appears to have a strong grip on South American markets.

Westinghouse has been the most aggressive vendor in terms of selling technology licences and its experience illustrates the dangers and advantages of licensing. In Japan (through Mitsubishi) and in France (through Framatome), Westinghouse has derived a stream of royalties and technological feedback. In Japan, they have also maintained a flow of hardware orders and are now collaborating with Mitsubishi in developing improved designs of reactor at a time when lack of home orders would not allow them to do so by themselves.

However, a third vendor, Kraftwerk Union which broke links with Westinghouse in 1970, and, more recently, Framatome have emerged as formidable competitors in world markets. For the future, unless Westinghouse can win orders in its own right, the collaboration with Mitsubishi may also turn against them with Mitsubishi becoming yet another competitor at Westinghouse's expense.

Company structure and strategy

Much of the previous discussion on technology transfer is paralleled in the structure of companies and the development of their capabilities. The capability and scope of a company can be considered in four dimensions, horizontal integration, vertical integration, product diversity and international scope. A fifth factor, ownership pattern, may condition some of the other factors.

A vertically integrated company is one which is involved at all stages between raw materials through intermediate goods to final goods, for example, in nuclear power, a vertically integrated company might have interests in uranium mining, fuel manufacture, power station construction and electricity sales. A horizontally integrated company has a high degree of self-sufficiency at a particular stage, for example a horizontally integrated nuclear power station vendor would be able to supply a high proportion of the components of the plant from its own factories.

The advantages of high levels of integration, particularly horizontally, are generally in greater control, whilst the disadvantages are in lower flexibility. A horizontally integrated company has control over prices and the availability of production capacity. However, in periods of low demand, their production facilities will lie idle and their capability may be lost. A company which subcontracts will be able to 'play the market' to obtain the best balance of price and quality and, in periods of low demand, has only design staff to carry.

Methods of production may vary according to the level of demand for the product. The normal categories used to describe the method of production of goods are one-off, batch and mass. It is expected that a larger scale of production will bring with it economies of scale (see later), that is, reduced unit cost, and it might be hoped that some quality benefits (notably consistency) would accrue. The adoption of mass-production or production-line methods also tends to carry the assumption that the production process is well understood, even routine.

With nuclear power, high levels of horizontal integration have generally been associated with production-line facilities. As was noted earlier the American vendors and more recently the French vendor have all tried to set up and maintain horizontally integrated production-line facilities (the capacity of such facilities is about 6 units per year and 'mass-production' would not be an appropriate description). German and Canadian reactors have, by contrast, generally been built on a batch basis and occasionally on a one-off basis, with a high degree of sub-contracting.

Product diversification, a company producing a range of distinct products, may give greater stability and may allow a larger pool of skills to be sustained than if the company was single mission. In addition, diversification may provide some insulation against short-term adverse market conditions when financial losses can be absorbed and skilled manpower redeployed. However, if the product is not central to the diversified company's plan, its commitment to the product may be less determined than a single mission company's would be.

Companies can be subject to a range of ownership patterns from, at one end of the spectrum, public ownership to investor-ownership at the other end. Public ownership may be at national, regional or local government level and within these categories control may be exercised directly through democratic assemblies or indirectly through government agencies such as the nuclear power research agency.

Often public ownership arises because the industry concerned provides a vital service or relates to a capability which is regarded as being strategically important. Public ownership is expected to give continuity and to partially insulate the service or the capability from the vagaries of market conditions. The main advantages to the company of public ownership are the ability to absorb short-term losses and sometimes to draw on the state's pool of personnel and research. Against these advantages, partial insulation against the effects of losses may lead to a lack of competitiveness and the company may be vulnerable to disruptive political direction particularly when governments change.

The strategic nature of nuclear power, particularly its military connotations, means that state-control has been a common, although not universal, feature of nuclear power. In addition, the scale of resources required to enter the field and the economic risks of such a lumpy (in terms of flow of orders) product has meant that investor-owned companies have been increasingly reluctant to commit themselves to this field.

Economic and technical improvement

For a piece of equipment such as a power station which produces a homogeneous product – electricity is not qualitatively different whether it is

produced by a hydro-electric power station or a nuclear power station – a manufacturer can make improvements to this equipment in three ways:
– changing the design or materials to reduce capital and/or running costs
– producing the same or similar design at lower unit costs (this is defined to include economies of scale)
– improving the reliability whilst still using the same basic design

In practice few changes fall precisely into only one category, but the distinction is a useful one for analytical purposes.

Changes in design or materials. A number of terms are used to describe changes or innovations in design and materials and these can be usefully ranked according to the scope of the change and the extent of the benefits. This hierarchy runs from product change through product development and technological change to technological progress. Technological changes are more significant than product changes and the terms 'development' and 'progress' carry an implication that the changes are in some sense beneficial. Change may be brought about in a series of small steps with successive designs changing only a little from their predecessor, known as incremental change, or in large steps with the new design bearing little resemblance to its predecessor, known as radical change.

Where a new product incorporates major innovations over its predecessor producing a 'step change' in economic performance, it is sometimes known as being of a new design generation or technology vintage.

Again, there are problems in applying this set of ideas to nuclear power plant, the most important of which are a result of the complexity of nuclear plant and the high degree of interdependence of the sub-systems. A nuclear power plant is the most complex product yet produced containing, for example, an order of magnitude more components than a spacecraft. Many of the sub-systems are highly complex technologies in their own right. Failure in any one sub-system is likely to render the whole system inoperable and may cause damage over the whole plant and beyond.

Identifying what constitutes a major technical change causes problems. Intuitively, the reactor core itself would seem to be the heart of the technology and thus only important changes to that should count as technological change/progress or as producing a new design generation. However, the reactor cores of the major reactor systems have proved extremely reliable and have seldom been an important determinant of reactor performance. There has been a great deal of detailed work 'fine-tuning' the operation of the core but in general the basic design of the core has not been the subject of much radical change.

This factor, the high cost (or even impossibility) of repair and the capital intensity of nuclear power plant (about two-thirds of the costs of nuclear electricity are fixed capital charges incurred whether the unit is in operation or not) has led to emphasis being placed on improving reliability and durability. This is in contrast to the focus for most other technologies including fossil-fired electricity generating technology where the emphasis lies with improving the basic operating parameters of the unit, for example for fossil-fired plant, its thermal efficiency, whilst maintaining reliability within acceptable limits.

Reducing unit costs. This can be further subdivided into economies of scale dealt with separately and a rather disparate collection of measures including improved scheduling, cheaper purchasing of parts, etc. This latter collection can be categorised as improved management and is difficult to place in any analytical framework although its impact can be important. Such effects are noted in the case study chapters.

Improving reliability. Good reliability is of particular importance to projects with high fixed costs (incurred whether or not the equipment is available for service) to ensure that the capital costs can be spread as thinly as possible. As with improved management in the previous category, it is difficult to place improving reliability in any generalised framework, but good reliability is a crucial aspect of the economics of nuclear power and the ways in which it has been achieved will be a major focus of the case study chapters.

Standardisation

The concept of standardisation has acquired relatively little prominence in most industries but in nuclear power it has assumed major importance. Even more than most of the terms discussed here, the concept of standardisation suffers from a lack of precision. At its lowest level, the term can be used to describe a technology which has the same appearance to users and which conforms to a minimum set of performance and safety criteria. For example, motor vehicles can be regarded as standardised to the extent that the foot pedals are in the same sequence, the lighting meets given standards in terms of type, performance, position and colour, and the brakes, stability and crash-protection meet minimum criteria. At this level, standardisation is relatively non-controversial acting as a safety net and having little impact on the design, manufacture and innovation process.

At its highest level, epitomised by the Model T Ford (any colour as long as it is black), standardisation requires not only that the design be identical but also that all components be identical being from the same source and manufacturing process. At this level, standardisation has the potential to significantly reduce costs but raises a number of major concerns relating to the risk of standardised error and the stifling of innovation.

The concern about standardisation in the nuclear power industry arose mainly from the seemingly perverse 'customisation' of the plant that was ordered in the USA during the second half of the 1960s (discussed in greater detail in Chapters 4 and 5). With four vendors offering NSSSs, 12 architect-engineers designing the balance of plant and over 50 utilities ordering plant to their own specification, it was inevitable that almost every plant, except for duplicate units on the same site, would be unique. This customisation did not extend to the NSSS which, from a very early stage, was subject only to incremental change, but occurred in the rest of the plant.

By 1970 it was clear that this had created appalling problems of regulation, with each plant requiring a separate safety analysis, and that many of the designs had severe defects delaying construction and escalating costs. The vendors responded by attempting to develop standardised nuclear islands (for example General Electric's 'GESSAR' design) and a number of utilities joined forces to commission a thoroughly standardised design (SNUPPS) from the architect-engineer Bechtel. However, these attempts were largely overtaken by the recession in ordering and the spate of cancellations which left no order surviving which had been placed after 1973.

In other countries, without the diversity of utilities, vendors and architect-engineers (of the four countries studied only USA uses specialist architect-engineers) and with the opportunity to learn from the USA's errors, the pressures for customisation have been very much less. France has opted for thorough-going standardisation on a 'Model T Ford' scale whilst Canada has used only three basic designs over its 22 commercial orders.

In FR Germany some customisation has been generated by the regulatory responsibilities which are partially devolved to the 11 *Länder* each of which may have different requirements. Attempts are now being made through the so-called 'convoy' system of licensing plant to obtain agreement at a federal level removing the diversity of requirements.

The pros and cons of standardisation (particularly to a high degree) are discussed in detail in subsequent chapters, but broadly the 'pro' arguments are that it:
- reduces regulatory workload as basic licensing approval can be given for a whole tranche of plants rather than on a one-by-one basis
- reduces costs by allowing batch or mass production of identical components. Similarly spares inventories can be minimised
- solutions to faults covering a number of plant need only be produced once and can then be applied to all plant

The main arguments against are:
- a serious design or materials error may be embodied in a large number of plants, potentially causing severe problems in maintaining electricity supplies if the plants must be shut down
- design improvements or innovations will tend to be stifled since their incorporation would violate the standardisation
- competitive pressures will be dampened if, in advance, it is known which design will be chosen especially if single sourcing of components is used

ECONOMIES OF SCALE

Since the 1960s the nuclear industry has argued strongly that nuclear power is a technology in which the economics should improve substantially as the size of unit increases. In practice, scale economies have been less evident in nuclear power than originally expected. Economies of scale can operate in a number of ways and, for analytical purposes, it is useful to split these according to the position in the product life-cycle at which they occur – during manufacture or during operation.

It is also useful to distinguish two, often conflicting, effects which can be termed 'economies of size' and 'economies of numbers'. Broadly economies of size arise because the marginal cost, at the production and operation stage, of producing and operating a larger capacity item is often less than the average cost or, to put it simply, doubling the size does not necessarily double the cost. Economies of number arise because, generally, the greater the numbers produced the lower the unit cost, partly because more efficient production methods can be used. Clearly a conflict between these two possible economies could arise if increasing the size of units significantly reduces the number required.

In general, the various types of scale economies are well known and understood, but there is less attention paid to reasons why expected economies of scale do not take place or even diseconomies occur. In the next section, which analyses scale economies according to their position in the life-cycle, careful attention is paid to these potential diseconomies.

Economies of scale during manufacture. As was discussed previously there are two, potentially conflicting, effects which can operate at this stage. Economies of size arise because increasing the capacity of a product generally does not require a commensurate increase in quantity of materials, for example a pint glass needs less than twice the amount of materials required by a half pint glass. In addition many items will vary comparatively little according to capacity, for example the control room for a 900 MW reactor may differ little from that required by a 1300 MW reactor. To set against these positive factors, fabricating larger items is often more difficult than fabricating smaller items. For example, large items, such as pressure vessels and turbine blades, are now reaching the stage where producing larger ones which would function reliably would present severe technical difficulties. If this increase in difficulty of fabrication is reflected in higher reject rates and higher costs during manufacture, or poorer operating reliability, the economies of scale are reduced.

Attempts to achieve economies of size were a particularly powerful influence on the direction of nuclear power technology development in the 1960s. In the USA, fierce competition between the four vendors led them to attempt to reduce unit costs by scaling up designs very rapidly. Many orders were placed for 1200 MW units three years before the first operating experience had been gained with units half that size. Elsewhere technologies were abandoned or given lower priority, notably gas-cooled reactors, because they seemed, for various reasons, less amenable to this type of scale economy.

As was discussed previously, methods of manufacture can be categorised as 'once-off', 'batch' and 'production-line' and increasing scale of production can produce economies of number.

Generally, production-line methods will be cheaper than batch methods which will in turn be cheaper than once-off. In addition the larger the throughput for a given method of manufacture, the lower the costs. This effect arises mostly because the overhead costs of setting up production facilities can be spread more thinly, but also because learning amongst the operatives should allow higher productivity and quality. However, it should be noted that the use of production-line techniques can only have a relatively limited effect on overall costs as many of the costs are incurred on-site in the civil engineering and assembly phases. Nevertheless, particularly if demand for the product is low, economies of number will be a significant effect and it may prove more economic to produce several units of a small capacity than fewer units of a larger capacity.

This trade-off is well illustrated by Ontario Hydro's decision to order their reactors in batches of four relatively small units and construct them on the same site. The alternative of building perhaps two large reactors was rejected on the grounds that the larger units would not be so reliable and any economies of size foregone would be balanced by economies of number achieved. Ontario Hydro stressed, particularly, the positive effects of having a stream of work available to its construction workforce which improved productivity and morale.

The practice of placing several units on the same site has been widely adopted and has a number of advantages (regardless of unit size) to the utility. Firstly, due to the difficulty in establishing new sites for nuclear plant, utilities have often used the same site, where a precedent for nuclear plant has been established, to house a number of units, sometimes duplicates, but sometimes of totally different reactor types. In such a situation, economies can be made on the cost of facilities such as access roads and transmission lines. Secondly, when reactors are built as identical twins, on the same site, it is often possible for them to share facilities such as control rooms, although, increasingly, safety considerations are dictating that units be built autonomously.

Economies of scale can also be achieved at the manufacturing stage by increasing the scale of raw materials purchasing – buying in bulk. Such benefits will be little influenced by whether a decision is taken to choose economies of size or number.

Economies of scale during operation. The major economies of scale during operation will arise from increased unit size. Historically larger unit sizes in fossil-fired thermal power plant have yielded benefits in manpower productivity and from economies associated with being able to buy and deliver fuel in larger quantities. For nuclear power both of these benefits will be rather small compared to other potential scale economies.

Diseconomies of scale during operation might arise due to electrical system considerations. To ensure a secure supply of electricity, the utility must have supply capacity in readiness to cover for the risk of break-down of the largest unit on the system. This cover is often supplied by maintaining thermal plant hot in readiness to start generation very quickly – so-called spinning reserve – and thus if a nuclear unit represents the largest unit on the system, costs of this cover will be increased. In addition, on system supply security grounds, it is generally considered sensible to ensure that no single unit represents more than about 10% of total system capacity. This may limit the size of unit that can be chosen by smaller utilities.

A further potential diseconomy of scale, complexity, is more difficult to define. Larger units will tend to require larger workforces, which may require more formal management procedures. It may be that small units on which the operating staff have much stronger personal links are able to operate with more informal and efficient procedures. Such an effect would tend to show up as improved reliability and reduced down-time. At the level of the equipment itself, larger units may require more complicated safety and ancillary equipment which may make the unit more difficult to operate, maintain and repair. Again such an effect might be difficult to demonstrate unequivocally but would show as poorer reliability and longer down-times.

SUMMARY

The standard methods of evaluating the development of a technology must be heavily qualified if they are to be of use in evaluating nuclear power. In particular, factors such as design complexity, long lead-times and difficulties of access to nuclear plant severely limit the ways in which technology can be developed. Nevertheless, the concepts of learning, technology transfer, user–producer relationships and standardisation appear likely to offer useful insights into the development of nuclear power.

Although none of the individual features which have made nuclear power development so difficult is unique to it – an accident at a chemical process plant could have an impact which would be felt over generations – the combination of these features is unique. Nevertheless many of the lessons learnt are likely to be applicable to other large-scale technologies such as chemical plant and space technology and possible future technologies such as fusion power.

CHAPTER 3

The economic evaluation of
nuclear power

Introduction

In this chapter we discuss the main determinants of the economic performance of nuclear power plant. In particular we examine the process of project
evaluation including the discount rate and electricity supply system
economics. Our benchmark in these discussions is the main competitor to
nuclear power generation, coal-fired generation. Following this general
discussion, the three specific parameters that are used in this book to
evaluate economic performance – operating performance, construction cost
and construction time – are introduced. The methods of calculating them
are given, their significance discussed and the potential pitfalls in comparing international experience noted.

The economic rationale of nuclear power

Compared to coal-fired generation, nuclear power has high capital costs
and low running costs. The simplest and most intuitively obvious way to
present nuclear economics is in terms of generation costs per kWh. Expressed in this way, capital-related items account for some two-thirds of
nuclear electricity costs, and operating expenses (mainly fuel) amount to
around one-third. The contrast with the main cost elements of the major
competitor to nuclear power – coal-fired electricity – is shown clearly in
Table 3.1. Coal-fired generation costs are predominantly fuel, and the
two-thirds to one-third proportion is almost exactly reversed. In general,
then, the relative competitiveness of nuclear power and coal depends on the
balance between the controllability of capital-related costs in the nuclear
case, and fuel costs in coal-firing. (The economics of coal-fired and nuclear
power stations were compared recently in Nuclear Energy Agency, 1983.)

Table 3.1. *Proportions of nuclear and coal generating costs represented by major cost components (%)*

	Nuclear		Coal-fired
	USA (1982)	UK (1983/84)	UK (1983/84)
Capital	69	66	28
Fuel	20	24	65
Other operating	11	10	7

Source: US Department of Energy, Energy Information Administration (1982). *Projected costs of electricity from nuclear and coal-fired power plants* DOE/EIA 035612, Washington DC, November; and CEGB *Statement of case* Volume 1 (1982). Sizewell B Power Station Public Inquiry, London.

The components of nuclear power costs

Nuclear power costs can be divided into five components:
– construction costs
– fuel costs
– other operating costs
– research, development and demonstration (RD&D) costs
– decommissioning costs

However, for reasons which are discussed later, only the first three sets of costs have had a major impact on economic assessments of nuclear power, despite the fact that issues connected with the other costs often loom large in public debates on nuclear power.

Construction costs. A breakdown of the estimated construction costs of a nuclear power plant is shown in Table 3.2. This breakdown is taken from the UK Central Electricity Generating Board's (CEGB) recent proposal to build a PWR at Sizewell. However, although this gives the general shape of costs, a number of factors means that there is substantial variation in such breakdowns between different projects. Such factors include:
– differences between reactor systems. For example, in the UK the nuclear steam supply system (NSSS) for an AGR was estimated to cost nearly four times as much per kilowatt as that required for a PWR (see Wilson, 1982, p. 117)
– differing costs of components. For example, in France where mass-production facilities have been set up, components may be much cheaper than in countries where such facilities do not exist

Table 3.2. *Typical breakdown of estimated nuclear power capital costs*

Item	Percentage of total cost
Civil engineering and building	21
Design and consultancy costs	15
Nuclear steam supply system	13
Turbine generator	12
Other electrical plant	9
Other mechanical plant	8
Contingencies	22
Total	100
Interest during construction	28

Source: Wilson, A. (1982) *CEGB Proof of Evidence No. 8* for the Sizewell B Power Station Public Inquiry, Vol. 2, p. 117, Central Electricity Generating Board, London, November 1982 and Sizewell B Public Inquiry, Transcript of Proceedings Day 10, p. 72.

– changes in costs through time. Changes in real labour and material costs may mean that the cost of individual items may change in real terms.

Equally important, however, such estimates have often not proved to be a very accurate guide to outturn costs, generally being underestimates. The reasons for this are complex and are analysed in detail in the subsequent chapters, but an important factor has been changes in design during the construction phase often requiring additional equipment. This has often not affected the major identifiable items such as the NSSS and the turbine generator but has increased the cost of most of the other items. Whilst the 'contingencies' item is specifically included to cover such changes, it has generally not proved adequate.

The other major item of expenditure associated with the construction phase is the interest incurred on the capital during this period, so-called interest during construction (IDC). This is dependent on the stage at which items are paid for and is also highly dependent on how close to schedule the plant is built. With lead-times frequently overrunning by several years, especially in the USA, increases in IDC have accentuated the rises in real construction costs.

Fuel costs. As noted earlier, fuel costs are a much smaller element of total costs than construction costs and, in recent years they have not proved prone to large, difficult to explain, increases. A typical breakdown of fuel costs, again taken from the Sizewell proposal, is given in Table 3.3, splitting

Table 3.3. *Analysis of fuel cycle costs*

Item	Percentage of total fuel costs
Uranium supply	53
Conversion	1
Enrichment	31
Fabrication	10
Waste fuel management	4
Total	100

Note: This set of data is taken from one of several scenarios put forward by the CEGB. In other scenarios, the weighting of each of the elements varies somewhat according to factors such as the forecast world price of uranium.
Source: Wright, J. K. (1982) *CEGB Proof of Evidence No. 9* for the Sizewell B Power Station Public Inquiry, Table 15. Central Electricity Generating Board, London, November 1982.

costs into five different elements, of which uranium supply and enrichment dominate. The largest element is uranium supply, and, after a sharp rise in costs in the mid-1970s, it now seems unlikely that this element of cost will increase rapidly in real terms.* The next largest element is enrichment; clearly this element would only apply to reactors which use enriched uranium such as PWRs and BWRs and not to reactors such as the Canadian CANDU which do not. As with uranium supply, there now appears likely to be an abundant supply of enrichment capacity available for the short- to medium-term and there are few grounds for anticipating sharp increases in this cost element.

Perhaps somewhat surprising is the low contribution of waste fuel management. A number of points should be made about this item. First, the CEGB evidence assumes that waste fuel is reprocessed and will produce uranium which has a positive value to offset some of the costs. Second, the costs given here are discounted at 5% real per annum (see later for a detailed account of the use and rationale of a discount rate). Since the CEGB assumes that reprocessing is not carried out until 40 years after removal from the reactor, this reduces the cost to only about 14% of the cost of reprocessing it today. Other countries have differing policies on the need for, and timing of, reprocessing. Third, reprocessing and waste disposal is not firmly established on a commercial scale and actual costs may turn out to be very different from these anticipated costs.

* For a detailed description of the factors underlying the evolution of uranium prices see Buckley *et al.* (1980).

Other operating costs. These costs are expected to be much smaller than the previous two sets of costs and cover a diverse set of elements including labour costs and repair and maintenance.

RD&D costs. RD&D costs of nuclear power have been (and continue to be) large in relation to other civilian technologies. By 1980 the US Government had committed $12.8bn to nuclear support (Hellman & Hellman, 1982, p. 157), and much official OECD expenditure – some $1.95bn in 1984 (International Energy Agency, 1985b, Table 5A, p. 39) – continues to go to 'advanced' reactor R&D (mainly breeders and fusion). The figures quoted in Table 3.1 take account of virtually none of this large state support, because they are based on utility calculations, and utilities have not generally financed significant nuclear R&D. On the other hand, as the R&D into current nuclear technologies is a sunk cost, there would be little point in making a notional retrospective charge on utilities or consumers at this stage.

Decommissioning costs. As with waste management costs, the projected costs of decommissioning are speculative but, due to discounting, have little impact on the economics of nuclear power almost regardless of what gross figure is projected for the costs of decommissioning.

PROJECT APPRAISAL

The costs detailed in the previous section clearly represent a major input to the appraisal of nuclear power projects, but a number of other elements are necessary in order to carry out a full appraisal. In particular it is necessary to see the nuclear power station as part of an electricity supply system. Its introduction will have implications for the whole system. The process of discounting costs, mentioned briefly previously, is also of crucial import-ance and in the following section the rationale and function of this process are examined.

Electricity supply system economics

Electricity supply systems are run in what is known as a 'merit order'. This means that plant available for service is brought on or shutdown to meet fluctuations in demand in order of marginal generation cost, that is, all things being equal, the cheaper the operating costs, the higher the utilisa-tion. This usually produces a merit order with hydro-electric power at the top (effectively zero marginal fuel costs) followed by nuclear power, solid fuels (lignite or coal) and oil/gas and within each of these categories newer plant is likely to be preferred to older, less efficient plant.* A further option is importing (or 'wheeling') power from neighbouring systems but the costs of this option cannot be generalised and will depend on what unused plant the neighbouring system has available to meet the extra demand.

It is important to note that the criterion for merit order operation is the marginal generation cost (primarily fuel) and takes no account of the fixed costs. Thus once a nuclear or hydro station has been built, it will almost always be run in preference to a coal-fired station. However, this does not imply the nuclear/hydro plant is cheaper *overall* than a coal-fired plant.

The effect of adding a new nuclear plant to an electricity generating system will be that, since it will be placed near the top of the merit order, all other things being equal, the utilisation of the stations beneath it in the merit order, predominantly fossil-fired plant, will be lower than it would have been without the nuclear plant. This will produce, by definition of merit order operation, savings in fuel costs. In an overall economic evaluation, these fuel cost savings must be offset against the capital charges associated with the nuclear power station.

The converse of these fuel cost savings is that whenever a nuclear plant is not available for service, for whatever reason, plant that would not otherwise be operating,† must be brought into service. By definition, this plant has higher operating costs than any plant then operating. These additional costs are known as replacement power costs and they are usually the major cost resulting from the shutdown of a nuclear power plant, generally far exceeding the direct costs of repair. An additional cost resulting from the non-availability of plant is that the capital charges (amortisation plus interest) must be spread over less output.

If, for example, the marginal cost of generation at a 1000 MW nuclear plant is 1 c/kWh and the marginal cost of replacement power is 8 c/kWh, the net cost to the utility for replacement power for the nuclear unit will be more than $1.7m per day. In practice, the cost of replacement power will depend on factors such as the time of day and the season – these will determine the level of demand and hence which plant is available to produce replacement power, and the geographic location – some areas have a cheaper mix of plants available. A utility will, as far as possible, schedule shutdowns over which it has some discretion at periods when replacement power is cheapest. Such shutdowns include those for refuelling, maintenance and non-urgent repairs.

---

* In practice, the merit order criterion does not always apply. Examples of this include when plant has to be run out of merit order due to transmission system weakness or because the marginal plant cannot be shutdown or started up quickly enough. Despite these exceptions, merit order operation is generally closely adhered to.
† An option might be to wheel power rather than use the utility's own plant.

Comparing these costs with the corresponding capital charges, a plant of 1000 MW might cost $2bn, generating annual capital charges (amortisation plus interest) of about $200m. In such a case, the loss of output through a plant breakdown would cost around $0.55m per day of capital charges not recovered, about a third of the replacement power costs.

Discounting

While at one level, the issue of discounting is a narrow and technical one, it is of particular importance to the economics of nuclear power. The idea, which lies behind discounting, that a benefit earned today is worth more than a benefit of the same monetary value earned in the future is intuitively reasonable. For example, if a sum of $100 available today could be invested to give $105 in one day's time, it should be given more value than $100 available in a year's time. The formal expression of this essentially simple idea is found in the practice of discounting, which is used in investment appraisal. In the example quoted above, our $105 should be discounted by 5% per annum to give a 'present value' of $100.

To carry out an investment appraisal using this procedure, so-called discounted cash-flow (DCF) analysis, all costs and benefits are discounted to give a 'net present value' of the project. A positive net present value suggests that the project will be profitable against the chosen discount rate criterion. If more than one use is available for the capital, this technique allows the choice of the most profitable use or if only one use is being considered, it establishes whether it meets a minimum rate of return criterion.

The reason why discount rates are so important in nuclear projects is that the present value of a nuclear investment is heavily dependent on the value of the discount rate chosen. In addition there is a great deal of potential discretion and even arbitrariness in the choice of this value. (Typical discount rates applied in the major nuclear power-using countries are tabulated in Nuclear Energy Agency, 1983, p. 32.)

Nuclear power has an unusual cost profile through time – very high initial costs, substantial net benefits during operating life, and a long 'tail' of waste disposal costs many decades into the future. By comparison, the main competition – coal-firing – has more modest initial costs and generating benefits and no tail. In the competition between the two, nuclear power is most favoured by 'middle range' discount rates. At high discount rates, the high short-term capital costs become dominant and the operating benefits become less significant. At very low rates, on the other hand, the combination of high capital costs and the long tail of waste management costs tends to swamp the operating benefits. Discount rate changes have profound implications for appraisal results. In the UK, for instance, the PWR for Sizewell is viable at the 5% rate used by the state utility (CEGB) but at the discount rates used by other state utilities (British Coal and British Gas both use 10%) or at the French Government's rate for Electricité de France (EdF) (9%), it would fail to win approval. It would also have difficulty in winning approval at discount rates around 1%. This latter result is to do with the spectacular differences that varying the discount rate can have over long periods of a century or more. This can be illustrated by looking at the present value of a cost of $1000 incurred in 100 years' time. At a 1% discount rate, the present value of this $1000 would be $370: this falls to only $8 at a 5% rate and to the virtually infinitesimal level of 7 *cents* at a 10% rate. From a cynical viewpoint opponents of nuclear power are faced with the interesting dilemma of whether to argue for very high or very low discount rates in their opposition to nuclear power: in practice either will tend to favour non-nuclear alternatives.

Setting discount rates

There are two basic ways in which actual discount rates are set. A discount rate can be set at a rate higher than the cost of borrowing the capital to act as a rationing device on that capital. Alternatively it can be set at the cost of borrowing the capital to reflect the actual costs associated with the project. This should ensure that all costs are recovered.

The discount rate as a capital rationing device. Applying a discount rate which is higher than the prevailing real interest rates is a device frequently adopted by governments to ration capital and ensure equity between the public sector and the private sector. In this case it is argued that the supply of investment capital is limited, and, because of the security of government debts, governments have privileged access to capital. Thus, if government projects are not to 'crowd out' more profitable investments in the private sector, governments should adopt some procedure whereby only those projects are adopted which achieve a rate of return comparable to those achieved by economically marginal projects in the private sector. This marginal rate of return is adopted as the discount rate and the government should proceed with projects which have a positive net present value. Private companies may also ration their capital expenditure in periods of cash shortage by using a high discount rate hurdle well above the cost of borrowing.

The discount rate as a reflection of the cost of capital. A more apparently neutral use of a discount rate is simply to apply the prospective cost of capital to the project. If only one project is under consideration, it can be established how profitable it is or if more than one is under consideration, the most productive use for the investment capital can be chosen.

In most countries, the practice is that future costs, benefits and interest rates are expressed in real terms, that is, net of inflation. However, this practice has not yet been adopted widely in the USA and the procedure needed for non-inflation adjusted costs and benefits requires that the rate of inflation be forecast. Only if constant inflation rates and gross interest rates are assumed, will these procedures produce the same result. In this chapter discount rates quoted will be real rates unless otherwise stated.

Project appraisal for electricity utilities

Generally the investment required for a new power station is an order of magnitude or more larger than any other single electricity utility investment decision. In addition, in the past, power station investment has been primarily required to meet growth in demand rather than replacement of existing capacity. In this situation the question becomes not whether to invest, but which option for power generation to choose. The perceived cost of not meeting demand is so high that any option is likely to meet the minimum criterion.

For the future, two factors mean that the option of not investing in new capacity will become much more realistic. These are:
- low electricity demand growth. If electricity demand growth remains considerably below historic levels, a far higher proportion of investment in power stations will be to replace existing stations. In this case, refurbishing existing stations becomes a significant option
- scope for electricity conservation. A viable option to make provision for future demand growth is now improving the efficiency of existing uses of electricity to such an extent that the projected electricity demand growth is counterbalanced by reductions in existing uses, obviating the need for new capacity

Actual discount rates

Nuclear investment is always carried out either by state-owned companies, or utilities subject to some degree of economic regulation and control. The issue of the appropriate discount rate to use cannot therefore be a straightforward application of market principles. However, the relationship between private sector discount rates and utility rates is extremely important, given the capital-intensity of nuclear (and other electric) investments. If utilities were to use lower discount rates than were applied in the rest of the economy, there would be a risk that investment capital would be misallocated being used less profitably than if they used comparable rates.

As was argued earlier, the use of project appraisal techniques on power station proposals has almost invariably been a question of which option to build rather than whether to build. In this situation, it makes little sense to use a high discount rate as a hurdle, and, particularly in the USA, private sector utilities tend to use market interest rates as the discount rate.

Because utilities generally have not been seen as being likely to go bankrupt, and have been assured of their ability to make a return on investments, utility investments have been seen as very low risk. This situation is changing, particularly in the USA, as is argued in detail in subsequent chapters. Economic regulators are now not automatically allowing a return on investments, and the possibility of utility bankruptcies is becoming more real. In this situation, the market is beginning to attach the sort of 'risk premium', frequently used in the private sector to reflect the risk of project failure to utility investments by reducing the 'bond rating' (this determines the cost of capital to a borrower) of utilities with plant construction programmes, particularly if they are for nuclear plant.

Whether this is an appropriate way to handle economic risk is arguable. Economists would argue that a sensitivity test, or some other explicit risk analysis, is more appropriate, but it is certainly an effective disincentive to nuclear investment in the USA.

In the public sector, as was argued earlier, the problem of misallocation is directly addressed by setting the discount rate at a level estimated to be the marginal private sector rate of return. This explicit treatment of the problem does not prevent considerable dispute about the level at which the discount rate should be set. Restrictions on government borrowing in many countries mean that many projects that clear the discount rate hurdle are not proceeded with because of further capital rationing. Some further, more opaque form of project selection is being undertaken, often to reflect political priorities.

Issues concerned with distant long-term costs

There are two distinct but related issues here: one is the appropriate method for discounting costs in projects with long-lasting costs and the other is how, in practice, to finance them. Both issues raise the question of equity between generations.

Most projects are not expected to give rise to significant costs or benefits beyond a 15 to 20 year time span, and in this period the assumptions underlying discounting are reasonable. Nuclear investment however gives rise to a stream of longer-term costs – unmatched by corresponding benefits

– hundreds of years into the future, particularly in the form of waste management. In these circumstances conventional discounting assumptions become highly questionable. In particular, the practice of discounting at a constant rate, say 5%, makes the implicit assumption that a 5% average rate of return will be available somewhere in the economy over decades into the future. The idea that a cost of $1000 incurred in the year 2087 can be valued at only $8 today depends on a continuous flow of investment opportunities at 5% per annum over a century. Such an assumption – though almost universal in nuclear project appraisals – is hardly prudent. It also has significant macro-economic implications if a large number of nuclear investments (as in France) have an impact on the national economy. For instance a set of costs which in 100 years' time would represent 1% of GDP, if the economy as a whole grew at 5% per annum, would represent 50% of GDP if the economy grew at only 1% per annum. Although in recent decades Western industrialised economies have grown rather faster than 1% per annum a longer historical view suggests that higher rates may be exceptional and that it would be extremely unwise to rely on high rates of economic growth over many decades into the future. Economists have debated for some considerable time how these considerations should enter into the practical choice of discount rates for projects with very long-term consequences. There is agreement only, however, on the theoretical inappropriateness of constant long-term discount rates, and not on the alternative strategy to follow. It is beyond the scope of this book to enter this area, except to say that there is clearly a powerful case to discount at low or zero rates beyond the time horizon when there can be reasonably well-founded predictions of rates of return and economic growth.

These issues are of course directly related to the question of how to finance long-term costs with its implications for inter-generational equity. There are basically two methods of financing future costs. Firstly, a 'sinking fund' could be set up at the time of first investment which would earn a return by investment and which would appreciate to cover the future cost. For example, applying a 5% discount rate, if an investment of $1000 would incur a further cost of $1000 forty years after initial investment, provision for this further cost could be made by setting up a sinking fund of $142 which would grow by 5% annually. Alternatively this future cost could be met from future revenue that would accrue as a result of this and other profitable investments.

In the former case the benefits of the investment are explicitly paid for by the 'generation' making the investment, provided the sinking fund can be invested to meet the rate of return required. In the latter case a later generation would appear to be paying for an earlier generation's benefits. However, it can be argued that these benefits may lead to further productive investments that will ultimately be enjoyed by the later generation. This later generation may therefore not necessarily be worse off than if the investment was not made.

Both methods are vulnerable to the criticism made of constant discount rates above. On either method, the later generation may end up with a much more burdensome cost if present rates of return or growth are not sustained. If it is thought that, for the sake of equity, the current generation should ensure that it pays for future costs, the size of sinking funds would have to be very much larger than is currently envisaged.

The key parameters

Because, as was argued earlier, the capital cost of the plant and the extent to which it can be utilised are dominant in establishing the success or otherwise of nuclear power plant, this book concentrates for its economic analysis on the parameters which determine construction costs and utilisation. These three parameters are operating performance, construction cost and construction time.

Operating performance

A good measure of the overall performance of a plant is its annual capacity factor (or, in UK parlance, load factor). This expresses the actual annual electrical output of a unit as a percentage of the output the plant would have produced had it operated uninterrupted throughout the year at its design rating. The capacity factor is a good indicator of the *economic* performance of a plant because nuclear plants have high capital costs and low operating costs (relative to coal-fired plant) and are thus planned to be operated to their maximum extent in order to spread the capital charges over as much output as possible. The capacity factor is also a good indicator of the *technological* performance of a plant unless the unit is not operated when it is available for service or it is operated at a lower output level than its capability. This situation is only likely to arise when an electrical supply system has more nuclear plant and hydro plant (hydro plant has lower running costs than nuclear plant and may be operated in preference to it) installed than is necessary to meet base-load. Up to the end of 1985, the only instance where a substantial number of plants have been affected in this way is France where the output of nuclear plants has had to be reduced outside the winter months in response to lower system demand. The thermal stresses on large components and potential problems with the fuel make utilities reluctant to regularly change the level of output quickly in a

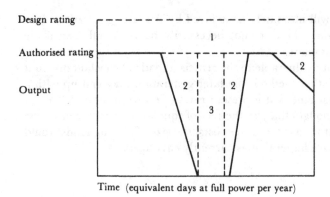

Time (equivalent days at full power per year)

Figure 3.1. Cause of lost output. Areas represent: 1, derating losses, 2, operating losses, 3, shutdown losses.

nuclear plant. This is an additional reason, therefore, why utilities are unlikely to operate plant at less than maximum capability.

In this book, analysis of plant performance is carried out at three levels of detail and over two differing time frames. At the most aggregated level of detail, simple capacity factors are examined. This analysis gives useful indications of the overall level of performance and may illustrate some aspects of learning, but it does not give any clues to the causes of any technical difficulties.

At the second level of detail, the analysis divides 'lost output' into three categories, derating losses, operating losses and shutdown losses (see Figure 3.1). All three categories are expressed as equivalent days at full power per year.

Derating losses. Capacity factors are calculated on the basis of the design rating of the plant since this is the rating on which economic appraisals are carried out and which reflects the expectations of the vendor. The authorised rating may differ from the design rating if it is judged that:
– it would be imprudent on safety grounds to operate the unit at its design rating
– operating the unit at its design rating would cause unacceptable wear on components
– more rarely, initial design conservatism or particular exigencies (e.g. the 1984 UK coal strike) may lead to later operation at higher than design rating

As a matter of convention, in the analyses in this book derating losses are calculated as being incurred whether the unit is in service or not.

Operating losses. These are incurred when the unit is producing power, but at a lower level than the authorised rating. This may arise for a number of reasons, of which the most important are:
- coastdowns. If a refuelling is due but system electricity demand is high, it may be more advantageous to the utility to run a unit at reduced power (the fuel is too depleted to reach full power) than to take it out of service for refuelling somewhat sooner. Such an operation is known as a coastdown
- minor equipment failures. Certain equipment failures may not necessitate a full shutdown but may require a reduction in output until repair takes place during a subsequent shutdown
- low electricity demand. As described above, if total electricity demand is less than the potential output available from hydro and nuclear resources, the output of nuclear units may be reduced below their maximum level. This is known as 'load-following'

Two figures are presented for operating losses, one, the unadjusted figure and the second, a normalised figure. The latter figure takes account of the fact that operating losses can only occur when the unit is in operation. This is because the importance of operating losses to a unit which is out of service for much of the year will be understated by the unadjusted figure.*

Shutdown losses. These are incurred when the unit has to be shut down completely: this is usually due to refuelling, equipment failure or maintenance and repair.

The third level of analysis expands these shutdown losses according to cause of shutdown and shows the shutdown frequency. Shutdowns are divided into 18 categories (see Table 3.4). The shutdown losses by cause are normalised to 8760 hours to reflect that, as for operating losses, shutdowns due to equipment failures can only occur when the unit is in service. The figures shown represent the number of hours that would have been lost for the given cause in achieving 8760 hours of actual operation. The shutdown frequency represents the number of times the unit would be shutdown in achieving 8760 hours of actual operation. This latter statistic gives an indication of the reliability of the unit.

It should be noted that the equipment failure rates do not necessarily reflect the extent of problems experienced with a particular plant system.

---

* The normalised figure normalises the derating losses to a base of 8760 hours of *actual* operation, for example if a unit operated for only half the year (4380 hours) and unadjusted operating losses were 876 equivalent hours at full power then the normalised operating losses would be 1752 equivalent hours at full power. In other words, on average when the unit in this example was producing power, it was operating at 80% of authorised rating.

Table 3.4. *Causes of shutdown nuclear plant*

===

Shutdowns are split into the following categories:
    A equipment failure, comprising failure in:
        1 reactor and accessories
        2 fuel
        3 reactor control system and instrumentation
        4 nuclear auxiliary and emergency systems
        5 main heat removal system
        6 steam generators
        7 feedwater, condenser and circulating water systems
        8 turbine generator system
        9 electrical power and supply system
        10 miscellaneous
    B operating error
    C refuelling
    D maintenance and repair
    E testing of plant systems/components
    F training and licensing
    G regulatory limitation
    H other

===

For example with some equipment problems, it may be possible to carry out remedial work during an otherwise scheduled shutdown such as for refuelling. For a utility, the most expensive problems are likely to be those that cause an unplanned shutdown, since replacement power costs will be highest. Repairs that can be carried out within, for example, a refuelling shutdown will be cheapest, although if the repair extends the period of shutdown the cost attributable to the fault rises steeply.

Analyses are split according to the age of the plant or according to the calendar year or time period. For example, analyses may show the performance of all plant of a given age or they may show the performance of plant in a particular year or time period. Analyses are divided up in this way to reflect the fact that some processes are time-related and that some are age-related. This distinction is discussed in greater detail in the previous chapter, but the main time-related processes include:

– overall technology learning. Learning may occur which will allow all similar units, regardless of age, to be operated more effectively

– the occurrence of specific incidents. For example the Three Mile Island accident may have caused additional or lengthier shutdowns at similar units, to ensure their safety and sometimes to retrofit additional safety systems

– the discovery of generic faults. The discovery of a previously unknown generic fault at one unit may make it prudent to shutdown and repair all similar units

In the latter category of age-related processes there are such factors as:
- site learning. As familiarity with the unit increases, the utility is likely to be able to operate and maintain it more effectively
- teething problems. When a unit is first operated, it is likely to suffer lower performance as installation errors are rectified and substandard components replaced
- wearing out failures. Some plant systems are likely to become less reliable as they age and wear out

To disentangle these effects thoroughly would require a large population of units such that, for example, the performance of similar units of five years' age in a given year could be compared with the performance of similar units of two years' age in the same year. Alternatively, they could be elucidated if there existed extremely simple mathematical proxies for these processes such that statistical techniques could be used to quantify these effects. In practice, neither condition is satisfied. Even in the USA, which has the largest operating history of nuclear plant, the former analysis is not possible.

Techniques such as regression analyses have been used to try to elucidate plant performance but the results have generally been unconvincing, explaining only a low proportion of the variability of plant performance, and proving unstable from year to year. It seems likely that this is due mainly to the large number of variables which could significantly affect plant performance but which are difficult to express mathematically. Such factors include the quality of the utility, the architect-engineering and the site construction work, and geographic considerations such as climate. In practice it is therefore necessary to interpret the analyses with caution, paying close attention to the specific details of the individual plants in the sample.

Construction time and construction costs

Construction time and construction costs are the remaining key parameters determining nuclear economics. Although in many respects, these two parameters are closely related – factors which tend to lengthen construction times also tend to raise construction costs – it is worth considering them separately as their impact on the economics and the problems of measurement are rather different.

Construction time. The impact of construction delays on nuclear economics is via two routes. The first is that delays result in the need for replacement power during the period of delay. These costs are rarely acknowledged, but delays lead to real and potentially very large replacement power costs. While many delays in recent years, particularly in the USA, have been deliberate 'stretchouts' often due to cash shortages, or low electricity demand growth,* the replacement power costs have still been significant. In comparisons of technical performance, therefore, it is important to identify stretchouts and take them into account when making comparisons with plant that has been constructed without such delays.

The second impact of construction delays is a reduction in the value of project benefits. Whenever a positive discount rate is applied, the deferral of benefits necessarily leads to a reduction in the present value of those benefits. The combined impact of the costs of replacement power and deferred benefits can be substantial: the CEGB calculated that a four-year delay in the completion of the Sizewell B PWR would lead to an increase in cost of £400m (Central Electricity Generating Board, 1982).

It is important to use common conventions in identifying at what point construction is regarded to have been started and at what point it is regarded as complete. For these purposes the required start date is when substantial site-work begins and in the analyses in this book this is taken as the date of the laying of first structural concrete. Other operations take place before this but they involve comparatively low expenditures and there is no very strong incentive to complete them quickly. The date of completion is taken to be the date of first commercial operation – this occurs some time after first power has been generated. A unit is declared commercial when it has completed its acceptance test – usually the test involves the unit running at continuous full-power for a specified period such as 72 hours. After this, the unit is handed over to the utility operating crew and is operated according to the utility's requirements.

Construction costs. The importance of construction costs is obvious enough: if capital charges represent two-thirds of the costs of a nuclear plant's electricity, a 10% increase in construction costs will lead to almost a 7% rise in electricity costs. Despite the importance of construction costs, it is difficult to make international comparisons of performance along this dimension. There are three main reasons for this. Firstly, currency exchange rates are unstable and rapid fluctuations of up to 20% or more are common. Such fluctuations could easily reverse an apparent advantage of one country over another. In addition, currency exchange rates may not be a good indicator of real values – they may be inflated by indigenous oil production or high real interest rates. Attempts to use purchasing power parities, a method by which the values of currencies are compared by

---

* In the USA, the economic regulators will not allow an economic return to be made on a power station unless it can be shown that it was needed to meet demand. This requirement is considered in detail in the chapters on the USA.

examining the costs of a 'basket' of goods and services, have not been entirely convincing.

Secondly, there are also difficulties in ensuring that construction costs are calculated using the same ground rules – for example, whether interest during construction has been included, whether research and decommissioning costs are included, what transmission facilities are assumed, whether the cost has been converted to constant money, net of inflation. In practice, few countries publish construction costs for which the ground rules are clearly identified and many costs used in studies of nuclear power economics are projected hypothetical costs rather than actual costs. Such hypothetical costs have been notoriously bad indicators of the actual costs incurred, usually proving to be a substantial underestimate.

Thirdly, even where a reliable time series of construction costs on a common basis can be derived, establishing which of the many possible influences on costs is actually contributing to any observed trends remains very difficult. Some of these influences include a change in the intrinsic work content of the plant (e.g. more man-hours or more materials), or a change in the level of disruptions to work (e.g. regulatory requirements, or financing difficulties). Such a disaggregation is only possible with considerable analysis on a plant by plant basis – impressionistic evidence may be misleading – and is beyond the scope of this book.

All this means that it is fruitless to attempt direct international comparisons of nuclear construction costs. What can be done is the practice followed in this book: namely to trace changes in construction cost *within* a country over time. This is itself not always entirely straightforward, as accounting conventions sometimes change, but it can be done and is much more reliable than direct international comparisons. It also allows a kind of second order international comparison in terms of difference in the rates of change of nuclear costs across countries. If, for instance, one country has stable construction costs while another exhibits rapid or persistent real (net of inflation) cost increases, it may be possible to draw some comparative lessons from such experience.

All three parameters – operating performance, construction time and construction cost – are clearly important as measures of economic and technical performance. However, differences in the nature and transparency of data available mean that operating performance is the area in which direct comparisons can most confidently be made. Reasonable comparisons can be made on the basis of construction times, but in the area of construction costs only limited and indirect inferences can be drawn.

CHAPTER 4

USA – Energy context and historical review

Introduction

The United States was the first country in the world in which a substantial programme of civil nuclear power reactors was undertaken. It pioneered the development of both types of light water reactor (LWR), the PWR and the BWR, and, through licensing and export arrangements, has transferred the technology worldwide. The roots of the technology are military – LWRs were developed as submarine propulsion units and used fuel enriched at plants which had, as their primary purpose, the provision of bomb-making material. Nevertheless, since the first *commercial* order for plant in 1963,* the sales of civil nuclear power plants have been under conditions rather closer to a free market than anywhere else in the world. There is normally a requirement for utilities to seek competitive bids, and there is a choice of suppliers. It has been estimated (Energy Information Administration, 1982) that government subsidies to the nuclear business had reached $13bn by 1980, but most of this expenditure had been on the fuel cycle and on advanced reactor designs such as the fast breeder, rather than on LWRs.

In the two chapters on the USA, one covering the historic development and the second assessing the economic performance and examining prospects, we look at the particular problems the USA faced as pioneer of the technology, how the early exposure of the technology to market forces affected its development and the effects of the absence of any centralised control.

One point to stress at the outset is that whilst the overall framework is common, the USA is a large and diverse country and there is no uniformity of experience – for almost every observation or rule, there is experience that directly contradicts it.

---

* The Oyster Creek order of 1963 was 'commercial' in the sense that it involved no overt subsidies and the order was placed in the belief that it was the cheapest option available to the utility for generating the power required.

Table 4.1. *Production and consumption of primary energy in the USA – 1973–84 (million tonnes of oil equivalent)*

	Production			Consumption			Consumption for electricity generation		
	1973	1978	1984	1973	1978	1984	1973	1978	1984
Solid fuels	381.4	429.9	560.1	361.0	419.2	499.3	221.7	267.7	356.9
Liquid fuels	486.6	491.3	497.8	770.1	894.8	715.2	84.2	87.7	29.6
Gas	514.5	450.3	417.7	526.7	470.1	430.2	85.6	75.1	74.2
Nuclear power	19.9	65.5	77.6	19.9	65.5	77.6	19.9	65.5	77.6
Other primary elec.ᵃ	62.7	64.0	74.4	62.7	64.0	74.4	62.7	64.0	74.4
Total	1465.1	1501.0	1627.6	1740.4	1913.6	1796.7	474.1	560.0	612.7

ᵃ Predominantly hydro-electric power.
Source: Energy balances of OECD countries, 1970–1982 and *1983–1984*, International Energy Agency, Paris, 1984 and 1986.

THE ENERGY AND ECONOMIC BACKGROUND

Because of its size and strength, the US economy has a dominant influence within the western industrialised world. Over the past two decades, the two oil shocks and the world recessions that followed them have had the most significant effect on the US economy, in both cases halting economic growth in the year or two following them. Thus GDP grew at an average annual rate of about 2.3% in the period 1973–84 compared to 3.4% in the period 1960–73.

The USA and the USSR are the largest producers and consumers of primary energy in the world, with the USA alone accounting for about 20% of world production and about 25% of world consumption. Since 1973, overall primary energy production has generally risen slowly and by 1984 was about 10% higher than in 1973. However, the changes in individual fuels were much more dramatic (see Table 4.1) with solid fuel production increasing by nearly 50% and natural gas production falling by nearly 20%. The contribution from nuclear power increased nearly four-fold. Energy consumption has fluctuated with the oil shocks and recession (see Table 4.2) and, overall, was little higher in 1984 than 1973. However, the breakdown of consumption between sectors had changed significantly with transportation and other use, primarily commercial use, increasing and industrial and residential use falling. Demand for fuel for electricity genera-tion increased by nearly 30% with much of this growth and the substitution of oil and gas used for electricity generation being taken up by coal. Nuclear

Table 4.2. *Delivered energy consumption in the USA by sector – 1973–84 (million tonnes of oil equivalent)*

	Industry			Transportation			Residential			Other			Total		
	1973	1978	1984	1973	1978	1984	1973	1978	1984	1973	1978	1984	1973	1978	1984
Solid fuels	115.5	113.3	103.0	0.1	0	0	14.4	21.1	25.0	0.2	0.3	8.5	130.1	134.7	136.5
Liquid fuels	81.1	100.4	77.3	408.2	471.2	450.3	100.5	73.6	38.8	55.6	63.8	67.7	645.4	709.0	634.1
Gas	183.2	140.2	131.9	0	0	0	116.1	130.2	108.8	61.7	54.4	69.7	361.0	324.8	310.4
Electricity	55.5	61.0	66.0	0.4	0.2	0.2	49.8	58.0	66.9	37.7	48.8	56.5	143.4	168.0	189.6
Total	435.3	414.9	378.2	408.7	471.4	450.5	280.8	282.9	239.5	155.2	167.3	202.4	1279.9	1336.4	1270.6

Source: See Table 4.1.

power made up about 13% of the primary energy used for electricity production. This proportion may rise to about 20% by 1990 although if electricity demand continues to grow, this proportion will fall again in the 1990s due to the absence of new orders for nuclear plant over the past decade.

In terms of consumption by final consumers (see Table 4.2), the USA has a number of features common to most industrialised countries:

- overall energy demand, after some fluctuations, fell slightly over the last decade
- electricity has increased its market share especially in the domestic sector
- the demand for oil is now falling in nearly all sectors

Unlike most other countries however, demand for natural gas is falling, perhaps reflecting a decline in confidence in the fuel in the late 1970s caused by fears about the future price and availability of gas.

THE DEVELOPMENT OF NUCLEAR POWER

The development of civil nuclear power in the USA can conveniently be divided into four phases.

The first, the 'atoms for peace' era, covers the period up to 1963 when the Oyster Creek order was placed. This was the first order to be placed for a nuclear plant by a US utility on the grounds that it was the cheapest option available to it. In fact, the vendor, General Electric, lost a great deal of money on Oyster Creek but this apparent demonstration of the fully-commercial status of nuclear power meant that no more direct government subsidies were available to LWRs. In the second (expansion) phase, from 1964 to 1974, the vast majority of orders for nuclear plants were placed (in fact all the surviving commercial orders). In the early years of this period, optimism about nuclear power was at its height, technological change was rapid and a large number of orders were placed. However, by the early 1970s this rapid expansion of nuclear power was beginning to cause concern, particularly on the part of the newly developed environmental lobby. The Calvert Cliffs decision of 1971, which required electric utilities to submit an environmental impact statement with applications to build new plant, was an important indicator of the much greater level of public scrutiny that utilities' decisions were coming under. In 1974 the Federal Atomic Energy Commission (AEC), the government agency responsible for both the promotion and regulation of nuclear energy, was split into the Energy Research and Development Administration (ERDA), a body which included the promotional role, and the Nuclear Regulatory Commission (NRC) whose major function was the regulation of nuclear facilities.

The third phase, from 1974 to 1979, marked the first period in which a substantial number of commercial plants were in operation and also the beginning of the decline in the fortunes of nuclear power. Whilst orders continued to be placed up till 1978, earlier orders were being cancelled; electricity demand growth was lower than expected and this required utilities to stretch-out construction schedules in order to delay plant until it was needed. Opposition to nuclear power became more sophisticated and the first major incidents with safety implications at large commercial plants occurred – notably the Browns Ferry cable tray fire and the Three Mile Island (TMI) accident. This latter event marked the end of the third period.

In the fourth phase, 1979 to the present day, fear of energy insecurity began to recede, and nuclear power lost much of its apparent strategic significance in combating energy dependence. Cancellations now occurred in plants where significant site work had been carried out. The spectre of utilities, once the safest investment on Wall Street, going bankrupt was raised and the date when further orders for nuclear power could be expected receded far into the future. Behind these events lay an increasing realisation that electricity demand growth would not regain its historic levels. Further, for new orders for generating plant, using standard accounting conventions, nuclear plant carried little if any advantage over coal-fired plant, but carried a much higher level of economic risk. It also became clear that the USA lagged behind much of the rest of the world in the quality of construction and operation of much of its plant. These failings had given intervenors ample scope to make real and substantial criticisms of the nuclear industry.

THE KEY INSTITUTIONS

The US institutional structure is complex, and many actors are involved at each level. Broadly the institutions can be divided into three groups: the electricity supply industry; the plant supply industry; and the economic and safety regulation. An additional influence, political decision-making, is more diverse and includes such factors as presidential initiatives, congressional influence on research budgets and the actions of other governmental (federal or state) bodies.

The electricity supply industry

The electricity supply industry comprises more than 2000 electric utilities which sell electricity both to final consumers and to other electric utilities (see Table 4.3 and Table 4.4). These utilities fall into three categories: investor-owned utilities, public utilities, and small co-operatives and municipally-owned utilities. Over 75% of total electricity supply is gener-

Table 4.3. *The 15 largest US power plant owners[a]*

Owner	Total capacity (MW)	Nuclear capacity (MW)	Ownership
1. US Government[b]	32 006	0	Federal
2. Tennessee Valley Authority	30 887	5 491	Federal
3. The Southern Company[c]	29 929	3 219	Private
4. American Electric Power[d]	20 740	2 154	Private
5. Commonwealth Edison	19 851	7 602	Private
6. Texas Utilities Electric[e]	18 593	0	Private
7. Middle South Utilities[f]	15 915	3 012	Private
8. Florida Power & Light	14 953	3 046	Private
9. Southern California Edison	14 681	2 610	Private
10. Duke Power	13 733	5 021	Private
11. Houston Lighting & Power	13 510	0	Private
12. Central and South West[g]	12 909	0	Private
13. General Public Utilities[h]	12 771	1 469	Private
14. Pacific Gas & Electric	11 857	1 084	Private
15. Virginia Electric Power	11 019	3 390	Private
	273 354	38 098	
Others	382 139	29 411	

[a]Capacity figures are as of June 1984.
[b]Comprises mainly hydro power owned by the US Corps of Engineers or the Water & Power Resources Service.
[c]Includes Alabama Power, Georgia Power, Gulf Power and Mississippi Power.
[d]Includes Indiana & Michigan Electric, Kentucky Power, Michigan Power, Ohio Power, Appalachian Power and Columbus & Southern Ohio Electric.
[e]Includes Dallas Power & Light, Texas Electric Services and Texas Power & Light.
[f]Includes Arkansas Power & Light, Louisiana Power & Light, New Orleans Public Service and Mississippi Power & Light.
[g]Includes Southwestern Electric Power, Public Service of Oklahoma, Central Power & Light and West Texas Utilities.
[h]Includes Jersey Central Power & Light, Metropolitan Edison and Pennsylvania Electric.
Source: Data on total capacity was supplied to the author by the US Department of Energy, Energy Information Administration. Data on nuclear capacity was drawn from the Science Policy Research Unit (SPRU) nuclear power plant data base.

ated by investor-owned utilities. Three of the largest of these, Commonwealth Edison (based in Illinois), Duke Power (North Carolina) and Pacific Gas & Electric (PG&E, Northern California) have frequently played a key role of technological leadership in electricity supply technology and have a substantial independent engineering capability. Other companies such as American Electric Power (AEP), Middle South Utilities and General Public Utilities own a comparable amount of plant through their ownership of a number of smaller utilities but do not possess a comparable engineering and research capability. In recent years, the migration of US population towards the south and south west has meant that utilities such as Southern

Table 4.4. *The major US nuclear utilities[a] (Nuclear Capacity, MW)*

Utility	Plant operating	Plant under construction	Total plant	Plant cancelled or suspended	Year of first commercial order
Commonwealth Edison	7 402	4 480	11 882	2 240	1965
Tennessee Valley Authority	5 491	4 800	10 291	9 912	1966
Duke Power	5 021	2 290	7 311	7 680	1966
Philadelphia Electric	2 130	2 130	4 260	2 320	1966
Georgia Power	1 561	2 200	3 761	2 226	1967
Virginia Electric Power	3 390	—	3 390	3 532	1966
Public Service Elec & Gas	2 205	1 067	3 272	5 667	1966
Carolina Power & Light	2 342	915	3 257	6 165	1966
Florida Power & Light	3 046	—	3 046	2 200	1965

[a]Capacities are as of January 1985.
Source: SPRU nuclear power plant data base.

California Edison, Florida Power & Light and Houston Lighting and Power have grown more quickly than other utilities.

There are two large public utilities, the Tennessee Valley Authority (TVA) and the Bonneville Power Authority (BPA). Both were originally set up mainly to exploit hydro potential although BPA does not actually own generating plant. TVA has for some time operated substantial quantities of thermal plant. Both sell electricity wholesale, mainly to municipal utilities and co-operatives. A substantial quantity of hydro plant is also publicly owned through the US Army Corps of Engineers and the Federal Water and Power Resources Services and the output from this is also sold wholesale.

The third category of utility is the co-operatives and municipally-owned utilities. These usually own very little plant but buy power from investor-owned or the large publicly-owned utilities.

This apparent fragmentation of the utility industry has led utilities to form a number of organisations which are intended to give them more coherence and technological strength. Two of these are long-established, the Electric Power Research Institute (EPRI) and Edison Electric Institute (EEI) whilst the third, the Institute of Nuclear Power Operations (INPO) was more recently established in 1979. EPRI, which was set up in 1972 and to which nearly all utilities subscribe, carries out and commissions research mainly on technical problems in power generation. Some of its research programmes, notably those aimed at nuclear power problems, have an international basis. EEI membership is restricted to investor-owned utilities and their aim is to represent these interests to government. It is also involved in training and acts as a forum for utility management. Unlike the other two organisations, INPO is specifically and solely concerned with

Table 4.5. *Westinghouse nuclear orders, commissionings & cancellations (MW (no. of units))*

	New orders	Percentage cancelled or indefinitely postponed	Plant commissioned	Plant cancelled[a]
1965	2 036 (3)	0		
1966	4 947 (6)	0		
1967	10 858 (13)	0		
1968	4 532 (4)	0		
1969	3 189 (3)	0		
1970	4 066 (4)	0	967 (2)	
1971	10 719 (11)	34	700 (1)	
1972	13 826 (12)	42	1 978 (3)	897 (1)
1973	17 993 (16)	67	3 051 (4)	
1974	7 996 (7)	100	2 987 (4)	5 676 (5)
1975	4 148 (4)	100	1 054 (1)	
1976			2 947 (3)	
1977			1 919 (2)	4 486 (4)
1978	2 240 (2)	100	2 007 (2)	7 448 (7)
1979				3 400 (3)
1980			907 (1)	4 350 (4)
1981			4 272 (4)	2 920 (3)
1982			1 148 (1)	
1983				
1984			3 268 (3)	4 373 (4)
1985			4 499 (4)	
1986			4 575 (4)	
1987			9 003 (8)	
1988			3 137 (3)	
1989			2 350 (2)	
Total	86 550 (85)	39	Commissioned 36 270 (39)	33 550 (31)
			Under construction 14 490 (13)	

[a] Does not include two units (Carroll County) ordered in 1978 but not currently being pursued (they do not have construction permits). It seems unlikely that they will be built, but they have not yet been formally cancelled.
Source: SPRU nuclear power plant data base.

nuclear power and its aim is to improve practice in all aspects of nuclear power plant construction and operation.

The power plant supply industry

The power plant supply industry can be subdivided into three sectors; vendors and main-component suppliers, architect-engineers and constructors. The four main plant vendors, i.e. those responsible for the design of the nuclear steam supply system (NSSS) are Westinghouse, Babcock & Wilcox (B&W) and Combustion Engineering (CE) which have supplied PWRs, and General Electric which has supplied BWRs (see Table 4.5 and Table

Table 4.6. *Babcock & Wilcox nuclear orders, commissionings & cancellations (MW (no. of units))*

	New orders	Percentage cancelled or indefinitely postponed	Plant commissioned	Plant cancelled[a]
1966	2 593 (3)	0		
1967	4 386 (5)	0		
1968	2 216 (3)	59		
1969				
1970	2 470 (2)	0		
1971	1 814 (2)	100		
1972	5 512 (5)	100		
1973	6 522 (6)	100	887 (1)	
1974	3 690 (3)	100	3 443 (4)	
1975			918 (1)	
1976	3 804 (3)	100		
1977			1 731 (2)	1 718 (2)
1978			906 (1)	3 450 (3)
1979				1 212 (1)
1980				7 781 (7)
1981				
1982				5 915 (5)
1983				
1984				1 310 (2)
1985				
1986				
1994			1 235 (1)	
1996			1 235 (1)	
Total	33 007 (32)	65	Commissioned 7 885 (9)	21 386 (20)
			Under construction 2 470 (2)	

[a] Does not include WPPS Unit No. 1 on which work has been suspended with the plant 63% complete. There are no firm plans to complete this unit.
Source: SPRU nuclear power plant data base.

4.8). Unlike vendors in Canada and FR Germany which subcontract main-component manufacture, all four companies have built up supply capabilities for many of the major components. However not all main nuclear components for a given order are necessarily supplied by the vendor; for example, the pressure vessel order for a Westinghouse sale may be placed with another vendor, e.g. CE or with a specialist company such as Chicago Bridge & Iron.

All four NSSS vendors are large diversified companies, although with respect to the power plant supply industry, their previous main contribution was either in supplying turbine generators – Westinghouse and GE – or in

Table 4.7. *Combustion Engineering nuclear orders, commissionings & cancellations (MW (no. of units))*

	New orders	Percentage cancelled or indefinitely postponed	Plant commissioned	Plant cancelled[a]
1966	1 283 (2)	0		
1967	4 215 (5)	0		
1968				
1969	1 070 (1)	100		
1970	4 237 (4)	0		
1971			805 (1)	
1972	3 670 (4)	77	825 (1)	1 620 (2)
1973	13 752 (11)	72	478 (1)	
1974	7 330 (6)	100		
1975			1 715 (2)	2 600 (2)
1976			830 (1)	
1977	5 040 (4)	100	845 (1)	
1978				1 836 (2)
1979				2 540 (2)
1980			912 (1)	3 570 (3)
1981				1 150 (1)
1982				7 642 (6)
1983			1 917 (2)	1 280 (1)
1984			1 087 (1)	2 570 (2)
1985			1 151 (1)	
1986			1 304 (1)	
1987			2 608 (2)	
Total	40 597 (37)	61	Commissioned 11 869 (13)	24 878 (21)
			Under construction 2 608 (2)	

[a] Does not include WPPS Unit No. 3 on which work has been suspended with the plant 76% complete. There are no firm plans to complete this unit.
Source: SPRU nuclear power plant data base.

steam supply systems – B&W and CE. During the 1970s they established the capability to supply the major nuclear components such as pressure vessels and steam generators from their own plants, although in recent years this supply capacity has been cut back. Smaller components have been purchased from a network of about 200 suppliers, sufficient to avoid a single-source supply of components, although if further orders are placed, this competitive bidding may no longer apply as more companies leave the field. Orders for the largest non-nuclear component, the turbine generator, are won by open tender and are not generally linked to the award of the contract for the supply of the NSSS.

Table 4.8. *General Electric nuclear orders, commissionings & cancellations (MW (no. of units))*

	New orders	Percentage cancelled or indefinitely postponed	Plant commissioned	Plant cancelled[a]
1963	1 270 (2)	0		
1964				
1965	2 124 (3)	0		
1966	7 691 (9)	0		
1967	6 248 (7)	10		
1968	8 346 (9)	27		
1969	2 944 (3)	64	1 270 (2)	
1970	2 940 (3)	0	794 (1)	
1971	4 583 (4)	52	1 999 (3)	
1972	16 501 (14)	72	2 762 (4)	2 230 (2)
1973	8 711 (8)	89		
1974	11 931 (10)	100	3 973 (4)	1 171 (1)
1975			4 022 (5)	4 718 (4)
1976				1 150 (1)
1977			1 886 (2)	4 610 (4)
1978				1 170 (1)
1979			784 (1)	2 400 (2)
1980				2 300 (2)
1981				1 711 (2)
1982			1 078 (1)	8 382 (7)
1983			1 011 (1)	3 504 (3)
1984			2 181 (2)	5 359 (5)
1985			3 326 (3)	
1986			934 (1)	
1987			6 249 (6)	
1988				
1989				
1990			1 065 (1)	
Total	73 289 (72)	53	Commissioned 26 020 (30)	38 705 (34)
			Under construction 7 314 (7)	

[a] Does not include Grand Gulf 2 on which work has been suspended with 22% of construction work complete. There are no firm plans to complete this plant.
Source: SPRU nuclear power plant data base.

The use of specialist architect-engineers (A-Es) (see Table 4.9) is a feature which distinguishes the USA from Canada, France and FR Germany. Essentially the function of the A-E is to co-ordinate and integrate the station's design details, and to assist in component specification and procurement. Some of the largest utilities, such as Duke Power and TVA carry out this function themselves, but smaller utilities without adequate technical resources for this function have used A-Es. Twelve specialist

Table 4.9. *The US architect-engineers*

Architect-engineer	No. of orders	No. cancelled or work suspended	Date of first order
Bechtel[+a,b,c]	61	26	1965
Stone & Webster[a,b]	28	16	1966
Sargent & Lundy[a]	21	7	1965
Ebasco[b]	15	8	1965
United Engineers & Constructors	13	8	1965
Gilbert[a]	7	1	1965
Southern Services	6	4	1969
Burns & Roe[a]	5	1	1967
Gibbs & Hill	4	1	1966
Offshore Power Systems[b]	4	4	1974
Fluor	3	0	1967
Black & Veatch	2	2	1973
Brown & Root[d]	2	0	1973
Unassigned	18	18	—
Utility	37	14	—
of which TVA	17	8	1966
Duke	13	6	1966
Others	7	0	—

[a] Architect-engineers with experience on pre-commercial LWRs.
[b] Architect-engineers which have won contracts as constructors.
[c] Bechtel has also participated in a number of other orders as joint architect-engineer. In other cases they have taken over the architect-engineering midway through the project.
[d] In 1982, Brown & Root was replaced as the architect-engineer for the two South Texas units by Bechtel.
Source: SPRU nuclear power plant data base.

companies have been involved in providing A-E services and the large ones such as Bechtel and Ebasco have wide-ranging interests in the design of other engineering and process plant such as offshore oil-production platforms and chemical plant. Utilities often have long-standing links with A-Es and this can restrict open competition in this market sector.

Constructors are responsible for supervising and scheduling on-site construction. Very often constructors also offer architect-engineering services although the A-E and constructor are often different for any given order. The existence of A-Es and constructors may, by lengthening the lines of communication, make the logistics of constructing nuclear power stations more complex than in Canada, France and FR Germany.

The regulators

There are two important categories of regulation, safety-related and economic. Safety regulation is now carried out by the NRC. Up to 1974, safety regulation was carried out by the AEC which was also responsible for the promotion of nuclear power and it became increasingly clear that these two roles were incompatible. This was resolved by splitting the AEC into the NRC and the ERDA, subsequently absorbed by the Department of Energy, whose role was widened to carry out R&D on all energy resources.

Economic regulation is much less centralised and is carried out by a number of bodies depending on the status of the utility. Retail sales by investor-owned utilities are regulated by state Public Utility (or Service) Commissions (PUCs or PSCs). The PUCs regulate all privately-owned public utilities including gas, water and telephones, and tariff changes have to be agreed with the PUC. PUCs generally have substantial research staff of their own, the ultimate decisions being taken by commissioners who are either elected or appointed, generally on a political basis. The number, term of office and method of appointment of commissioners vary from state to state. Long-term wholesale rates for electricity traded between states are regulated by the Federal Energy Regulatory Commission (FERC) whose commissioners are also political appointees.

Political influences

The federal government is less directly involved in commercial nuclear power, particularly recently, than in other countries, though it commits funds to advanced reactor development and to fuel cycle R&D. In the pre-commercial phase, the AEC was involved in promoting LWRs by providing subsidies for prototype plant but, increasingly, the government's role has been limited to providing fuel cycle services, especially enrichment, and to influencing the climate in which nuclear power operates, directly by the appointments it makes to such bodies as the NRC, and indirectly by its public pronouncements.

THE 'ATOMS FOR PEACE' ERA (1946–63)*

This period is bounded by the passage of the McMahon–Douglas Bill as the Atomic Energy Act of 1946 and the placing of the order for the Oyster Creek plant in 1963. The McMahon–Douglas Bill was intended to open up nuclear energy to the civilian sector through the AEC, but military considerations led to a concentration of effort on plutonium production and the development of submarine propulsion units. The Act largely restricted development of nuclear power to government agencies. However, by

* The material for this section is mostly drawn from the following sources: Dawson, 1976; Roth, 1982; Mullenbach, 1963; Perry *et al.*, 1977.

December 1953, President Eisenhower, in his 'atoms for peace' speech to the UN, began to encourage international development of power reactors by proposing the sale of key materials such as enriched uranium to overseas countries. Moves to open up nuclear power in the domestic market to private competitive participation culminated in 1954 in a new Atomic Energy Act. This allowed private companies to own nuclear facilities and the AEC to give grants to private enterprise for R&D and prototype plant construction although not for commercial facilities.

The first civil reactor for electricity generation, the Shippingport reactor, was already on order by this time and the 1954 Act led to the setting up of the Power Reactor Demonstration Programme (PRDP) in which the AEC subsidised a number of utilities building demonstration plants. Although a number of reactor designs other than LWRs were proposed, of the 14 units that were finished, nine were LWRs (see Table 4.10).

The two vendors who subsequently dominated reactor sales, Westinghouse and GE, had been at the heart of nuclear developments since the 1940s through the design and manufacture of submarine propulsion units. Even before the Oyster Creek order, both of these companies had already built up strong licensing links in Europe and Japan which were to form the basis of most of their export orders (see Table 4.11).

The AEC had put substantial sums of money into the research laboratories of both companies and they were well placed to win the subsidised orders of the PRDP. Along with B&W who had supplied the reactor (also an LWR) for the first nuclear-powered surface ship, the NS Savannah, only Westinghouse and GE won orders for commercial-sized units (150 MW or larger) in this period. The architect-engineers, Bechtel, Stone & Webster and Sargent & Lundy gained valuable experience with LWRs by their participation in the design and construction of these units.

By 1963, disillusionment at the apparently slow pace of development of civil nuclear power was becoming apparent. However, in December 1963, this disappeared with the award of the first commercial (unsubsidised) order for a civil nuclear plant. Jersey Central Power & Light awarded the contract for the Oyster Creek plant to GE, which was to supply a 650 MW BWR on a turnkey (fixed-price) basis. GE and Westinghouse severely underestimated the costs of large units and both firms apparently suffered substantial losses on the turnkey contracts which they undertook in the year or two after the Oyster Creek award. Nevertheless Oyster Creek marked the end of an era and had two important consequences.

First, it marked the end of further government financial support for LWR construction. Since nuclear power now appeared to be competitive with coal, it was argued that further subsidies were not necessary. Government money on reactor construction has subsequently been channelled into

Table 4.10. *Power reactors ordered in the pre-commercial phase*[a]

Station[b]	Technology[c]	Electrical output (MW(e))	Year ordered	Year commissioned	Owner/operators[d]	Vendor[e]	Year of final shutdown
Shippingport*	PWR	60	1953	1957	AEC/Duquesne	West	1974[f]
Indian Point 1	PWR	265	1955	1962	Con Ed	B&W	1974
Dresden 1	BWR	200	1955	1960	Comm Ed	GE	1978
Yankee Rowe*	PWR	175	1956	1960	YAEC	West	–
Humboldt Bay	BWR	65	1958	1963	PG&E	GE	1976
Peach Bottom 1*	HTGR	40	1958	1967	PEC	GA	1974
Big Rock Point*	BWR	72	1959	1963	Con Pwr	GE	–
La Crosse*	BWR	50	1962	1969	DPC	A-C	–
Haddam Neck*	PWR	582	1962	1968	CYAPC	West	–
San Onofre 1	PWR	436	1963	1968	So Cal Ed	West	–

[a] Includes power reactors with an electrical output greater than 40 MW(e) that were still operating in 1970.

[b] Plants marked with asterisks are those which received subsidies under the Power Reactor Demonstration Programme.

[c] Technologies are: PWR, pressurised water cooled and moderated reactor
  BWR, boiling water cooled and moderated reactor
  HTGR, helium cooled, graphite moderated reactor

[d] Owners/operators: AEC, US Atomic Energy Commission
  Duquesne, Duquesne Power and Light (Pennsylvania)
  Con Ed, Consolidated Edison (New York)
  Comm Ed, Commonwealth Edison (Illinois)
  YAEC, Yankee Atomic Energy Company (New England)
  PG&E, Pacific Gas & Electric (North California)
  Con Pwr, Consumers Power (Michigan)
  DPC, Dairyland Power Co-operative (Wisconsin)
  CYAPC, Connecticut Yankee Atomic Power Company
  So Cal Ed, Southern California Edison Company
  PEC, Philadelphia Electric Company

[e] Vendors: West, Westinghouse
  B&W, Babcock and Wilcox
  A-C, Allis Chalmers
  GA, General Atomic

[f] Shippingport was shut down in 1974, modified and recommenced operation in 1977 as a light-water breeder reactor before being finally shutdown in 1982.

Source: SPRU nuclear power plant data base.

Table 4.11. *US nuclear exports*[a]

Country	Vendors	Orders capacity (no. of units)	Year of first order
Belgium	Westinghouse licensee:		
	– Acecowen	2 931 (4)	1968
Brazil	Westinghouse	684 (1)	1971
France	Westinghouse licensee:		
	– Framatome[b]	41 940 (43)	1961
FR Germany	General Electric	250 (1)	1962
	General Electric licensee:		
	– AEG/KWU	7 702 (9)	1964
	Westinghouse licensee:		
	– Siemens/KWU[c]	2 167 (3)	1964
	Babcock & Wilcox licensee:		
	– BBR	1 308 (1)	1973
India	General Electric	420 (2)	1964
Italy	General Electric	160 (1)	1958
	General Electric licensee:		
	– AMN	2 912 (3)	1968
	Westinghouse	270 (1)	1956
Japan	General Electric	3 801 (5)	1965
	General Electric licensees:		
	– Hitachi	4 544 (5)	1970
	– Toshiba	8 972 (10)	1969
	Westinghouse	4 016 (5)	1966
	Westinghouse licensee:		
	– Mitsubishi	7 692 (9)	1970
South Korea	Westinghouse	5 137 (6)	1969
Mexico	General Electric	1 350 (2)	1969
Philippines	Westinghouse	651 (1)	1974
Spain	General Electric[d]	1 434 (2)	1965
	Westinghouse[d]	4 862 (6)	1965
Sweden	Westinghouse	2 760 (3)	1968
Switzerland	General Electric	3 475 (4)	1966
	Westinghouse	728 (2)	1965
Taiwan	General Electric	3 242 (4)	1969
	Westinghouse	1 900 (2)	1974
Yugoslavia	Westinghouse	632 (1)	1973

[a] This table includes direct exports by US vendors and plant built by licensees in their home market up to the end of 1985. It does not include sales by licensees to third countries.
[b] Includes only orders placed before Framatome terminated its licence with Westinghouse in 1981.
[c] Includes only orders placed before KWU terminated its licence with Westinghouse in 1970.
[d] A number of orders for Spain have been cancelled and are not included in this table.
Source: SPRU nuclear power plant data base.

high-temperature reactors and breeder reactors, neither of which routes has yet led to an order for a commercial plant. (Although several orders were placed for commercial-size high temperature reactors in 1971–74 these were all cancelled in 1975.) Second, the strong competition between GE and Westinghouse to reduce costs and take a large share of the orders placed meant that almost all of GE and Westinghouse nuclear resources were concentrated in improving the economics of LWRs. This was to be achieved mainly through economies of scale and reducing materials costs.

How the LWR came into a position to scoop this first massive pool of orders is a complex subject and a detailed discussion is beyond the scope of this book. Nevertheless it is worth summarising the main reasons. These are that:

- technologies other than LWRs stood little chance of winning orders against the apparently proven merits of the LWR. No plant larger than 100 MW using any other technology had been completed
- both on its internal research effort and its funding of the subsidised Power Reactor Demonstration Program in the 1950s, the AEC devoted the bulk of its human and financial resources to LWRs, especially to PWRs
- the key vendors, Westinghouse and GE, had earlier experience with LWRs from military contracts, which meant that they were wary about committing large resources to new concepts. Few other potential vendors had the technical resources to bring an independent reactor design to fruition
- the existence of enrichment capacity surplus to military requirements meant that enriched uranium was available very cheaply and that reactors which could use unenriched uranium did not appear as attractive as they might have done
- LWRs were thought to have some advantages over their competitors, particularly an expected capital cost advantage resulting from relatively high power densities and consequent small physical volumes.

THE EXPANSION PERIOD (1963–74)*

The Oyster Creek order was the most obvious public sign of the intense competition between GE and Westinghouse to open up the large market for power stations to nuclear power. It was the first of the so-called 'turnkey' orders for which the vendor guaranteed the price of the plant to the utility allowing for inflation.† In all, 12 turnkey orders were placed (seven to

---

* The material for this section is mostly drawn from the following sources: Dawson, 1976; Roth, 1982; Perry *et al.*, 1977; Bupp & Derian, 1981.

† The San Onofre plant, ordered in 1963, but on which the Westinghouse offer was made in 1959, was described by the utility as 'fixed price but not turnkey'. It is sometimes counted as a 13th turnkey order.

General Electric and five to Westinghouse) with most of the orders in 1965 being won on this basis. Although it is now clear that the vendors lost substantial sums on all these orders, the acceptance by vendors of such contracts had the desired effect on utilities. This was to overcome the utilities' reluctance to commit capital to projects that would otherwise have been economically risky.

By 1966, when 20 orders were placed, the vendors had become aware that their turnkey contracts were underpriced and that the costs of plant were difficult to predict. This meant that turnkey contracts were a substantial risk and the vendors withdrew their offer. Nevertheless, they were still able to draw ten utilities with no previous experience of building and operating nuclear plant (including the influential TVA and Duke Power) into placing orders for such plants.

The two largest makers of conventional steam boilers, Babcock & Wilcox and Combustion Engineering, won their first commercial orders for their own designs of PWR. Babcock & Wilcox had designed the early proto-typical Indian Point 1 (265 MW) plant which used an oil-fired steam superheater and a thorium–uranium fuel cycle. However, for their subsequent orders their designs were much closer to those of Westinghouse. Combustion Engineering PWRs were also similar to those of Westinghouse. They did not have the benefit of supplying small units prior to the 1966–68 burst of orders and their first order (Palisades) was for a large unit with a capacity of 805 MW. The consistently poor performance of this unit, due in part to a number of design faults, probably reflects their inexperience.

The existence of substantial scale economies was largely unquestioned and six of the 20 orders placed in 1966 were for units larger than 1000 MW, a full two years before an LWR of even half that size had entered service.

At this time General Electric was offering plant at a slightly lower cost than Westinghouse although both companies were winning a roughly equal share of orders. More importantly, real costs for both vendors were rising sharply. The cost per kilowatt of the turnkey plants was a little over $100/kW, but within two to three years, costs of new contracts had more than doubled. This reflected not only the underestimation of costs in the turnkey contracts but also continuing real increases in costs. This was in contrast to expectations that 'learning' would substantially reduce capital costs. In addition there was a strong expectation that economies of scale, as plant size increased from about 600 MW to 1000 MW and larger, would be considerable. This proved not to be the case and economies of scale were at best small (see later for a discussion of the factors behind this). Reactor orders declined from their peak of 31 in 1967 to only seven in 1969 as the industry absorbed this first wave of orders.

By 1970, fossil-fuel prices had begun to increase more quickly and nuclear orders began to pick up again, led initially by utilities with existing programmes. However, in the peak years (1973–74) when 79 orders were placed, 13 utilities with no previous experience of nuclear power placed orders. Technologically, new developments were still taking place. General Atomic, which was near to completing the prototype Fort St Vrain plant, won its first commercial orders in 1971 for its helium cooled graphite moderated design. Westinghouse won orders for plant which was to be almost entirely factory built and then towed to its site on a barge – it was hoped that this would overcome problems of construction site control which were at that time blamed for cost and time overruns. The orders arising from both of these developments have all been cancelled.

Despite the aura of success for nuclear power that this flow of orders gave, the political climate for nuclear power had become much more constraining by the early 1970s compared to that surrounding the first wave of orders in the mid-1960s. The environmentalist movement, which rose to prominence in the mid-1960s, was initially concerned with the ecological impacts from routine operation of facilities. In this respect, nuclear plants which operate at a somewhat lower thermal efficiency than fossil-fired stations have a slight disadvantage in terms of heat releases, but an enormous advantage in terms of emissions of acid rain precursors and carbon dioxide. Later, however, attention switched from routine operation to the consequences of accident sequences, especially core melt-downs. Three decisions in the early 1970s illustrate these new constraints.

First, intervenors challenged the AEC's licensing of the Calvert Cliff project and in 1971 won a judgement that the AEC was not exempt from the provisions of the National Environmental Protection Act: it therefore had to consider the total environmental impact of the station including the thermal releases to the environment. The result of this was a delay of about 18 months in all reactor licensing and a requirement that all applicants produce an Environmental Impact Statement (EIS). At about the same time a debate on the standards for emergency core cooling systems (ECCSs) arose leading to the issuing by the AEC of 'interim acceptance criteria'. These were widely challenged and an 18 month 'rule-making' hearing ensued. This resulted in the publication of revised criteria in 1973 (which also did not satisfy objectors). Finally, in 1973, the AEC began to actively encourage standardised plant as a means of reducing an increasingly heavy licensing workload. At the same time the AEC limited unit size to 3800 MW thermal or approximately 1300 MWe.

These three decisions illustrate the pressure and difficulties facing the AEC in the early 1970s. The task of regulating the construction of more than 200 nuclear reactors, almost all of which were customised 'one-offs', was exacerbated by decisions such as that relating to Calvert Cliff which increased the scope of licensing activity.

The conflict of interest inherent in the role of the AEC as both regulator and promoter of nuclear power was resolved shortly after the Arab oil embargo by the decision to split the AEC into the NRC and ERDA. ERDA, which was subsequently absorbed into the Department of Energy, inherited all the AEC's development activities and, in response to pressures for greater efforts to be concentrated on new energy forms, added similar responsibilities for the other energy supply technologies, such as solar power and coal conversion technologies.

Despite the new constraints, the outlook for nuclear power still appeared reasonably favourable at the end of this period for a number of reasons:

– the economic ground that nuclear power had lost due to capital cost escalation appeared to have been recovered by the quadrupling of oil prices and the subsequent impact on coal and natural gas prices
– nuclear power with its development through the fast breeder reactor was not seen as resource (i.e. uranium) constrained and could thus be seen as a secure source of energy for a long-term future unlike the fossil fuels
– the reactors ordered in the mid-1960s were beginning to come on stream and to supply significant amounts of power

This optimism was to prove unfounded – 1974 effectively marked the end of nuclear ordering in the USA, with all subsequent orders being cancelled.

FIRST COMMERCIAL OPERATION 1974–79

The commercial orders placed after 1966 were beginning to come on stream by 1974 and from 1973 to 1976, 31 units entered commercial service. For the first time it became possible to compare actual performance in terms of capital costs, and operating performance with expectations.* The record was disappointing (see the next chapter for a detailed analysis of economic performance) but the analyses were not accepted by nuclear power's supporters and decisions on plant ordering were, at that time, little affected by this.

Of much greater importance were lower than expected electricity demand growth and a concurrent cash shortage amongst utilities, together with additional costs and delays as the NRC tried to come to terms with evaluating the wide variety of designs facing it. Any optimism that the oil crisis had opened up new opportunities for nuclear power was short-lived and whilst a trickle of orders was placed between 1975 and 1978 (13 in all) no substantial progress was made on any of them before they were cancelled or indefinitely postponed. In any event, even these new orders were more

* The pioneering work in this area was carried out by Bupp & Derian (1981) and Komanoff (1976).

than balanced by cancellations and in 1974–75, 19 orders were cancelled; eight of these were for helium cooled reactors, resulting from the withdrawal of General Atomic from all its nuclear contracts.

In addition, the most serious safety-related incident at a large commercial plant up to that time occurred at Browns Ferry in 1975. This bizarre event was caused when an electrician accidentally set fire to a cable tray with a lighted candle which he was using to carry out a leakage test. The ensuing fire disabled the safety systems on the two operating units at the site (a third was not yet complete) and put both units out of operation for 18 months. Whilst this incident had considerable repercussions on the perceived safety of nuclear plant, the economic implications were perhaps not fully assimilated. This was partly because the plant was reparable (unlike TMI) but also because the plant was operated by TVA, which, as a large federally-owned utility not subject to PUC regulation, was able to absorb the costs much more readily than would have been possible for a smaller private utility.

The Three Mile Island accident in 1979 occurred when, for a variety of reasons, the feedwater system supplying the steam generators failed causing the core to be partially uncovered. The factors behind this accident were complex and are fully described in the report of the commission set up by President Carter to investigate the accident (the so-called Kemeny Commission) (President's Commission, 1979). The accident left the affected plant (Unit 2) irreparably damaged and the problems of 'cleaning-up' the plant are still to be resolved. Unit 1, although unaffected by the accident, only re-entered service some seven years later after much concerted and determined opposition.

As well as these conspicuous incidents, a number of generic problems that have continued to plague LWRs were beginning to emerge. Steam generators in PWRs were subject to corrosion leading to repeated shutdowns for repairs, whilst for BWRs, the problems were more diverse including leaking valves and seals as well as the pipe-cracking which has subsequently become more apparent (see the next chapter for a more detailed description of these problems).

By the late 1970s, the temporary falling off in demand growth began to look more long-term in character. Utilities, which had slowed or halted work at construction sites because of low demand growth and/or cash flow problems, found that increasing the tempo at such sites was not as simple as was expected. Site productivity problems, changing regulatory requirements, escalating real equipment costs and the high cost of finance meant that such 'stretchouts' proved very costly. This was reflected in another spate of cancellations during 1977–78 when a further 23 units were cancelled.

Overseas, the export markets the USA had begun to build up (such as

South Korea and Spain) were not living up to expectations, neither were new countries placing orders for plant. This was for three main reasons. First, electricity demand growth was falling short of expectations, second, non-US suppliers were beginning to win orders and finally, finance was proving difficult for such capital-intensive projects especially for less developed countries.

The nuclear industry, including utilities, tended to attribute these problems to factors which were mostly external to the industry, notably:
– the world recession, for depressing electricity demand growth
– the high cost of borrowing, for escalating costs
– the NRC, for not providing a stable regulatory background
– intervenors, for delaying the completion of plant

Little attention was paid to the problems internal to the industry itself. There was still an expectation that further orders for nuclear plant were just round the corner. The combined effects of the TMI accident and the second oil crisis soon changed this perception and the priority in the nuclear industry changed from being the maintenance of a flow of work to keeping intact critical resources so that some minimal capability to build nuclear plant could be retained.

THE POST-TMI PHASE

Whilst in terms of public health effects TMI was not a significant event, in financial and public relations terms it was a disaster. One of the factors that made the accident so significant was that, whereas at Browns Ferry the accident was clearly the result of a human error combined with inadequate fire protection (which could be readily improved), at TMI few of those associated with the plant could escape from responsibility. The operators, utility, vendor, architect-engineer and NRC all attracted some significant element of blame and it was clear that a whole range of remedial measures was necessary which went well beyond a few simple hardware 'fixes' which would ensure that a similar set of events could not happen. Whilst the responses to TMI were in some cases immediate, for example the loss in public confidence, some were slower and linked to other developments. For example the adverse reaction of financial institutions to nuclear construction programmes was also due to the continued low growth in electricity demand. The adjustments that have followed TMI can be divided into three categories: financial effects, utility reforms and regulatory reforms.

Financial effects

The psychological shock to the financial community that an asset with a replacement value of billions of dollars could be written off in minutes, leaving an even larger clean-up bill and effectively rendering the other unit on the site inoperable for a long period, was profound. At the same time, a combination of rapidly increasing fuel prices and the world recession that followed the second oil shock led to low or even negative electricity demand growth leaving utilities very short of cash. Utilities with large plant construction programmes, especially for nuclear plant, were particularly hard-hit. Financial markets generally reduced the bond rating of such utilities (and hence increased the cost of borrowing). Utilities, faced with a cash shortage, increased cost of construction (partly through increased cost of borrowing and partly through the effect of regulatory backfits) and apparently little demand for new plant, were obliged to severely reduce their construction programmes. Between the time of the TMI accident and the end of 1984, 64 plants were cancelled including, for the first time, units on which significant construction work had been carried out. Whilst large utilities such as Duke Power and TVA managed to cut their programmes in half in a reasonably orderly fashion, cancellations and construction problems associated with some smaller utilities highlighted many of the underlying problems. This was especially so for the Washington Public Power Supply System (WPPSS), and Cincinnati Gas and Electric which cancelled the Zimmer plant when it was said to be 97% complete.

The WPPSS scheme was conceived in the early 1970s by WPPSS and BPA and was underwritten by many local municipalities and co-ops. Eventually it grew to a 5-unit project involving three different vendors and three different A-Es. Almost all contracts for each plant were open to competitive bidding and the result was a complex web of suppliers and contractors which the project management was unable to control. In 1980, the NRC ordered work to be stopped on unit 2 (the furthest advanced in construction terms) pending changes in the management structure. These were carried out and seem to have had some effect (this unit is now in service).* The changes were too late however to save the other four units which fell victim to low growth in electricity demand and financing difficulties. In 1981, units 4 and 5 were mothballed (24% and 3% complete respectively) and in 1983 work on the remaining units, 1 and 3, was also stopped with 63% and 76% respectively of construction complete. WPPSS defaulted on its debt and the financial situation of the co-ops and municipalities involved will take a long time to resolve.

The Zimmer plant, owned by Cincinnati Gas & Electric, suffered many quality control problems including lack of documentation and even falsifi-

---

* Note that electricity demand in this region has fallen so far short of expectations that not even this single unit can be maintained on base-load and must be shut-down during some of the spring and summer months.

cation of records. Whether the plant was poorly constructed, or whether there was insufficient documentation to prove that it was well constructed, is not clear. The problems however, were such that when the plant was virtually complete, the NRC issued a stop-work order. An audit by Bechtel suggested that the additional cost of completion would be about $1.5bn. The plant was cancelled and may now be converted to burn coal.

While spectacular, these instances are not isolated: other examples, such as the Shoreham, South Texas, Midland and Grand Gulf projects, also illustrate a combination of poor management control and severe financial problems.

By 1981/2, rising fuel costs and difficulties of funding construction projects had left the finances of most utilities in an unprecedently poor state, and the spectre of 'rate shock' had been raised. Rate shock would occur if a utility built a power station which would significantly expand its asset base – perhaps by more than 50%. Most PUCs do not allow the utility to incorporate the asset in its rate base which allows it to earn a return on investment until it is producing power and has been judged a prudent investment. In some cases the scale of nuclear investments in relation to existing utility assets has raised fears of 80% tariff increases when the plant is completed. The apparently dire financial condition of many utilities led to a great deal of debate about measures designed to improve utility cash flow such as allowing utilities to begin to recover cost before the plant is complete, the so-called principle of allowing 'construction work in progress' (CWIP). By 1984, increased electricity demand resulting from the recovery of the US economy and more stable fuel prices meant that, with only a very few exceptions utility finances were improving. Delays and cancellations meant that the impact of rate shock on consumers was limited. Nevertheless the experience of this period is such that a utility facing an apparent shortage of power in the future would now be most unlikely to turn to nuclear power because of its uncertain and open-ended financial implications.

Utility reforms

Whilst there are US utilities which are well-managed and which have built and operated nuclear power plant to the highest standards, the TMI accident and the managerial failings that led to the problems at plants such as Zimmer and Midland, have illustrated that the regulatory climate and public opinion will inevitably be conditioned by the weakest utilities. In simple terms, an error by one utility could cost every utility dearly. This had led to debates in some quarters about the utility arrangements for future nuclear plant – for example, whether their use should be restricted to large utilities. More practically, the Institute of Nuclear Power Operations (INPO) was set up.

The rationale for INPO was that many of the problems of escalating capital costs, more stringent regulatory requirements, poor operating performance and the generally poor public image of nuclear power could be ameliorated simply by ensuring that good construction and operating practices were diffused throughout the industry.

The first president of INPO (Dennis Wilkinson) was appointed from outside the utility community – his background was in the nuclear sub-marine fleet. He chose to limit his objectives initially doing only the most important programmes first to ensure that those programmes undertaken were performed thoroughly. Four programmes were set up at the outset:

- inspection of operating plants by task force teams, which identified weaknesses and, through careful follow-ups, ensured their rectification
- review and analysis of any abnormal events, and dissemination to utilities of any significant information arising from these reviews
- setting training standards for all personnel involved in operation, and maintenance
- providing support to improve emergency planning

Subsequently, five further programmes have been set up, the most important of which are the accreditation of utility training programmes and the inspection of plant under construction. In addition, utilities which felt they had particular problems have been able to approach INPO for special assistance. By early 1980, INPO had attained 100% membership of all US utilities operating and/or building nuclear plant. Later INPO expanded to include international plant vendors and utilities from outside the USA, including the UK CEGB and EdF of France.

These programmes are rooted in the recommendations of the Presidential Commission (1979) set up to review the TMI accident, the Kemeny Commission. Above all, they stress the importance of disseminating data on all aspects of nuclear plant construction and operation amongst utilities as widely and effectively as possible.

INPO has been able to ensure that its recommendations are carried out by having two forms of pressure in reserve. First, arising from the TMI accident, it was deemed necessary to increase the insurance cover of nuclear-owning utilities. A consortium of insurance companies (NEAL) bears the risk over $500m and it is now a condition of insurance for utilities that they provide NEAL with copies of INPO's evaluation reports and information on the utility's corrective actions. Thus INPO can exert considerable pressure on utilities if they are reluctant to comply with INPO's recommendations.

Second, it was perceived that the visibility of nuclear power in utility management needed to be raised and that peer pressure would be an effective means of ensuring utilities' compliance. Thus, Wilkinson himself was part of many of the inspection teams including the first priority, the initial evaluation of operating plant. He enlisted top executives from other utilities to accompany these teams. This ensured the attention of top management at the utility under review.

Concurrently, and sometimes in combination, EPRI was bringing to fruition a number of programmes, such as those on PWR steam generator problems and BWR pipe-work cracking, which were aimed at remedying the major operating problems with existing plant. Recently, its focus has become more forward-looking with the launch of a new programme aimed at defining the detailed performance specifications for future plant. The focus will be on 'debugging' existing technology rather than on radical change.

The significance of these initiatives will be discussed in greater detail in the next chapter but, overall, they may indicate a recognition of the importance of utilities being informed and responsible customers for nuclear power plant.

Regulatory reforms

The TMI accident damaged the credibility of the NRC very seriously and the Kemeny Commission found a great deal to criticise about the operation of the agency. Whilst some of the recommendations could lead (at least conceptually) to well-defined actions and reforms, others were much more open-ended and potentially contradictory.

In the former category were:
- redesigning the control room. The TMI control room was found to be confusing – 100 alarms were triggered in the early stages of the accident, key warnings were not visible to the operators and several instruments went off-scale – and did not embody modern electronic technology
- improving training. The Kemeny Commission found many deficiencies in training, poor communications between vendor, utility, and A-E, and an over-emphasis on the consequences of single failures
- improving communications. The Kemeny Commission was critical of the lack of any consistent means of communicating important operating information between utilities and vendors. This was particularly relevant to the TMI accident as an incident at another B&W plant, Davis Besse, had strong similarities to the precursors of the TMI accident. If these had been acted upon, the TMI accident might have been avoided
- improving emergency plans. It was clear to all that TMI exposed a lack of emergency planning and response. There was considerable confusion

about roles and responsibilities which exacerbated the anxiety of local residents

However, many of the other findings and recommendations were much less clear-cut and easy to act on. The TMI accident seemed to cast doubt on theoretical precepts on which the Rasmussen Report of 1975 was based (United States Atomic Energy Commission, 1975). This report, which underpinned the industry view that nuclear power was adequately safe, assumed that the probability of two independent failures was negligibly low whereas this actually occurred at TMI (the Browns Ferry accident was also, in Rasmussen's terms, not 'credible'). The Kemeny Commission found that NRC had paid inordinate attention to 'worst case' type accidents and too little to less dramatic failures. This opened up recognition of the possibility of a vast array of minor equipment failures which, in combination, could have serious consequences.

The Kemeny Commission also highlighted the pre-occupation with the regulations and the danger that utilities would believe that simple compliance with regulations would ensure safety. It also criticised the complexity of regulations and the lack of attention paid by utilities and regulators to human factors.

Perhaps most difficult to deal with was the evidence the Kemeny Commission found that the NRC was still influenced by the old promotional philosophy of its AEC days and tended to 'err on the side of the industry's convenience'. As evidence for this, it criticised the reluctance of the NRC to apply backfits to operating plant. Kemeny was also highly critical of the NRC's management structure, particularly the ill-defined role of the five commissioners. The Kemeny Commission found a lack of co-ordination, communication and direction amongst the various departments and recommended a thorough restructuring of NRC.

This last recommendation, which the NRC could not by itself effect, has not been carried out and without this, the other recommendations have been difficult to implement. The NRC was already at the time of TMI under severe pressure from the industry to reduce and rationalise the number of backfits, yet carrying out Kemeny's recommendations has multiplied the number of backfits required. Increasingly it is felt within the industry that, whilst in isolation each backfit makes sense, in combination the effect on safety may be marginal or even deleterious. This could arise in a number of ways. One of the consequences of the strong vendor competition of the late 1960s was a drive to reduce costs. This manifested itself in smaller and smaller containments culminating in the Westinghouse 'ice-condenser' design. What space there was for maintenance and access has been significantly encroached upon by backfitted systems. It is also felt that the systems effects of many backfits have not been thoroughly evaluated –

backfits may duplicate functions or in combination frustrate their desired effect.

The NRC has tried to pay more attention to human factors and to less dramatic equipment failures but, in doing so, it has proved difficult not to exacerbate the problems of overcomplex regulations and lack of co-ordination as a whole new range of issues are opened up.

Overall, in the period since the TMI accident, the whole industry has made considerable efforts to improve its performance but many of the problems are deep-seated and, as discussed in the next chapter, may require more fundamental changes.

CHAPTER 5

USA – Assessment and future prospects

Introduction

The previous chapter described the institutions active in the nuclear industry and discussed the key events in the commercial history of nuclear power in the USA. This chapter is divided into three sections. The first analyses the economic and technical performance of commercial plant, including operating performance, capital costs and construction times; the second examines how the institutional structure has affected performance; and the third looks at some of the institutional reforms currently being proposed and how effective they might be in reviving the option to order nuclear plant.

ECONOMIC PERFORMANCE

Operating performance

The USA has far more operating experience with nuclear plant than any other country in the world with approximately 40% of the reactor years of commercial experience that have been accumulated in the WOCA, more than twice that of any other country. However, whereas in a country such as France the reactors are very similar and can be treated largely as a homogeneous sample, in the USA there are a great number of variables all of which could have an important influence on operating performance.

Table 5.1 shows the year-by-year performance of US nuclear units since 1972. These figures exclude pre-Oyster Creek units all of which must be regarded as prototypical and which are not really representative of current designs. It also excludes units which are in the first or second calendar year of commercial operation since 'new' units are likely to suffer from teething problems such that their performance in the first two years may not be representative of that in the long term.

Table 5.1. *Performahce of mature commercial units*[a]

	Capacity factor[b] (%)	Net capacity (MW)	No. of units
1972	68.7	1 270	2
1973	68.7	3 031	5
1974	53.3	6 558	10
1975	57.6	10 507	16
1976	62.7	16 023	23
1977	64.7	26 895	36
1978	68.7	34 614	45
1979	60.9	38 363	49
1980	57.3	43 838	55
1981	57.4	47 657	59
1982	55.9	48 441	60
1983	55.1	50 260	62
1984	55.7	54 532	66
1985	59.0	55 680	67
1986	56.3	58 674	70

[a] Includes only units ordered in 1963 or after (post Oyster-Creek) and units with more than two calendar years' service.
[b] Capacity factors are weighted by unit size to give greater weight to larger units.
Source: SPRU nuclear power plant data base.

This group seems to present a fairly coherent picture of a steady upward learning curve from 1974 to 1978 with a severe decline from 1979 (the year of the TMI accident) onwards. Overall, the yearly averages, even in the peak year of 1978, are far below the levels expected when the orders were placed in the mid-1960s, when capacity factors in excess of 80% were anticipated. However, the crude averages conceal a great deal of variability. Even a more detailed examination of performance which distinguishes different technologies, vendors, size ranges etc. and gives indications of the causes of poor reactor performance, does not adequately explain this variability, although it does provide useful insights.

However, prior to such a detailed analysis of the technical performance of each vendor's reactors, two more general observations which shed some light on this variability are worth making. These concern the geographical pattern of operating performance and the performance of duplicate units.

If the mean capacity factors after year 2 of post-Oyster Creek reactors with four or more years commercial operation are calculated and reactors are sorted into descending order of capacity factor, a very striking geographical pattern emerges. If the top 20% of reactors (i.e. the top 13) are considered, all six reactors in the adjacent mid-west states of Wisconsin and

Minnesota are to be found in this group and three out of five reactors in New England also appear (the two pre-Oyster Creek reactors situated in New England also have excellent records). In both areas a number of utilities is involved and so this effect cannot simply be explained as being the result of the excellence of a particular utility.

By contrast, no units from the South, the West Coast and the South East Coast appear in this group. Although the composition of this group of the top 20% of reactors is not random in terms of unit sizes and vendors – it is dominated by Westinghouse and General Electric units of less than 700 MW and Combustion Engineering units – the fact that similar units from outside these regions do not appear in the top group suggests that the geographical location may be significant.

The second factor is the performance of duplicate units. Almost without exception, the performance of the second or subsequent unit is very similar to that of the first unit, be it very poor as in the case of the Brunswick units, or very good as in the case of the Point Beach or Prairie Island units.

These two observations suggest that features specific to any given unit are crucial to its long-term performance. Such features might include:
- the quality of the utility management that is associated with the unit
- the design of the balance of plant and the integration of the NSSS into the overall design, that is, the architect-engineering
- the management qualities of the architect-engineering and constructor teams – even within the same architect-engineering company there may be considerable variability between teams
- the level of skill in, and the competition for the local pool of labour both for construction and operational personnel

These factors are difficult to quantify or even to establish firmly but they do suggest that any theories about performance that rely too heavily on technological factors such as who the vendor was, or whether the unit had suffered from generic faults, should be regarded with caution. For example, the Westinghouse Point Beach units have suffered severe steam generator corrosion culminating in the replacement of the steam generators at one unit and very extensive 'sleeving' at the other. Despite this, both units have maintained high capacity factors and unit 2 (the one that has been sleeved) has the highest long-term capacity factor of any US unit. Thus whilst a good basic reactor design and an absence of generic faults is desirable, it is by no means the only factor in determining reactor performance. We return to the significance of the geographical pattern later.

Nevertheless some temporal trends and vendor, size and learning effects do emerge on closer examination of the data and these are discussed in the next section. The size ranges which are used to analyse performance are

Table 5.2. *Performance of mature Westinghouse units by time period*

	2-loop		3-loop		4-loop	
Period	Capacity factor (%)	No. of reactor years	Capacity factor (%)	No. of reactor years	Capacity factor (%)	No. of reactor years
1976–78	79.3	16	67.5	15	64.7	8
1979–81	74.6	18	46.6	21	58.6	21
1982–85	77.3	24	64.3	37	54.9	39
1986	82.3	6	69.5	10	51.8	12[a]

[a] Due to a serious lack of confidence in the TVA's nuclear plant management, NRC and TVA agreed in August 1985 that all TVA's nuclear plant should be shut down. No firm date for the plants' restart has yet been set. The five affected units include two 4-loop Westinghouse PWRs (Sequoyah 1 and 2). If these units are discounted, the average for mature Westinghouse 4-loop units in 1986 rises to 62.2%.
Source: SPRU nuclear power plant data base.

most conveniently derived from the Westinghouse design which is modular in concept. Different sizes of unit are obtained by changing the size of the core and altering the number of coolant loops. Thus a 2-loop unit yields approximately 600 MW, a 3-loop unit about 900 MW and a 4-loop unit about 1200 MW. GE's design is not modular in this way but the size ranges defined by the Westinghouse design – less than 700 MW, 700–950 MW and larger than 950 MW represent a useful subdivision. The units supplied by B&W and Combustion Engineering are predominantly in the middle size range and are most sensibly treated as one sample.

Westinghouse units

The performance of US Westinghouse units is summarised in Table 5.2 and Table 5.3 and shown in greater detail in the Appendix to this chapter. Table 5.2 shows the mean capacity factor by 3 or 4-year time periods for mature units (for these purposes defined as units with more than two years commercial service). This table identifies any temporal trends such as technology maturation and the impact of specific events such as the TMI accident. The inclusion of only mature units should avoid distortions that might occur if a large number of units entered service in one period (this might be expected to lower the mean capacity factor below its long-term level). Table 5.3 shows mean capacity factors by age of unit and should illustrate the extent of unit maturation as teething problems are overcome and utilities learn how to operate the unit most effectively.

Table 5.3. *Performance of mature Westinghouse units by unit age*

Age	2-loop		3-loop		4-loop	
	Capacity factor (%)	No. of reactor years	Capacity factor (%)	No. of reactor years	Capacity factor (%)	No. of reactor years
1	63.1	6	61.4	11	62.1	19
2	67.7	6	60.5	11	48.7	14
3 onwards	76.8	71	61.3	87	56.4	80

Source: SPRU nuclear power plant data base.

Although there appears to be a reduction in performance with increasing size and, in addition, each size category (particularly 2-loop units) has a distinctive performance profile, there are a number of features which are common to all sizes of unit, which are examined below.

Steam generator corrosion. Steam generators have been a serious problem for Westinghouse units. They have been the cause of a significant number of equipment failures particularly in older units and have contributed to lengthened maintenance and repair shutdowns. Of the 10 units that entered commercial service before 1974, five have had to undergo complete steam generator replacement (an operation that has taken up to 15 months) and most of the others have required extensive repairs by sleeving or plugging.

This corrosion has taken a number of forms:

– denting. This problem arises from degradation of the support plates for the steam generator tubes. Corrosion of these support plates causes crushing and denting of the steam generator tubes. It is now thought that denting can be avoided in new units by using stainless steel rather than carbon steel for tube supports, and by controlling the water chemistry much more rigorously (avoiding seawater contamination of the secondary side – see below)

– wastage. This involves corrosion on the secondary side of the tube and appears to have resulted from the use of a high level of phosphates to control water chemistry. From 1975 onwards, new units have used 'all volatile treatment' (AVT) and most existing units also switched to AVT. These measures appear to be effective in reducing wastage.

– crevice corrosion. This appears to result from the existence of long crevices between the tube and the tubesheet which have become the focus for corrosion. Crevice corrosion should not occur if such crevices are avoided during manufacture

- stress corrosion cracking. This has occurred at highly stressed sections of the steam generators such as short-radius U-bends
- pitting. This is a local corrosive attack which is associated with the presence of high concentrations of chloride or oxygen
- vibration. This factor is perhaps less tangible than the others but may be equally important. Any corrosion process will tend to be exacerbated if, in normal operation, the steam generator has a tendency to vibrate. It can only be avoided by careful detailed design and installation

There are a number of other less important mechanisms which may be avoided if different materials or heat treatments are used. Overall, it appears that, with the current state of knowledge, the rate of corrosion that led to the premature ageing of some units should not now occur but that, for existing units, steam generator corrosion is likely to be a continuing problem, and steam generator replacement operations will be a feature for some time. Whether steam generator corrosion will cease to be an important issue for new units remains to be seen.

A problem that has arisen in recent years concerns a particular design of steam generators, the 'D'-type, which embodies a pre-heater. The pre-heater was incorporated to produce a steam generator of the same physical proportions as earlier models but with a higher steam (and hence electrical) output. The first unit with this design to enter service was in Sweden (Ringhals 3). It was found, after several months of operation, that severe corrosion of the steam generators had occurred. Subsequently it became apparent that this was due to excessive vibration caused by an error in design. The completion of a number of US units was delayed pending the development of methods for reducing the scale of the problem and one unit, McGuire I, which had entered service before the problem had been identified, was forced to run at only 50% output for about a year before a 'fix', which took about a month to carry out, had been developed. A further generation of steam generators, the 'E'-type, also embodied this fault but all units were modified before entering service. The latest vintage of steam generator, the 'F'-type, which entered production in 1976 and is fitted in units now entering service does not utilise a pre-heater and has not suffered from this acute corrosion problem.

This design fault was financially serious, delaying the completion of a number of units which caused utilities substantial replacement power costs and requiring considerable effort to develop a 'fix'. However, because much of the repair work was carried out before the units entered commercial service, the seriousness of the error is not fully reflected in the capacity factor data.

Method of cooling. For those units which use salt water as the coolant, rather than fresh water or cooling towers, a particular problem has been corrosion of the condenser tubes where inadequately corrosion-resistant materials were used. This required repairs to these condensers and has led to contamination of the water in the secondary circuit which may have accelerated steam generator corrosion. This problem is now reasonably well understood and has been controlled by backfitting, where necessary, corrosion-resistant condenser tubes and ensuring more careful control of the water chemistry on the secondary side. Note that this problem is not restricted to Westinghouse units, although the consequences, particularly for BWRs, may be different. The extent of the problem may be masked since its most serious impact appears as leaks in the steam generator tubes. Replacement of the condenser tubes can be carried out during scheduled shutdowns and thus will tend to appear as a somewhat lengthened refuelling shutdown.

Teething problems. Equipment failures in the main heat removal system, the feedwater condenser and circulating water system, the turbine generator, and the electrical power and supply system decline rapidly after the first two years of service, as do shutdowns for testing of plant systems/components. A further illustration of unit maturation is shown by the lower level of shutdown frequency and operating losses as units get older. Equipment failures in the reactor and accessories, fuel, nuclear auxiliary and emergency system, and the reactor control system and instrumentation have not been a significant cause of lost output, nor have shutdowns for training and licensing.

TMI impact. Although there was a sharp decline in overall capacity factors in 1979, the year of the TMI accident – the mean for mature Westinghouse units was 17 percentage points lower in that year than in 1978 – a more detailed examination reveals a rather different picture. 2-loop units were little affected, 3-loop units were severely affected by the onset of extensive steam generator repairs and only 4-loop units were significantly affected by regulatory restrictions. This suggests that, in general, the substantial backfitting that has been required in the wake of the TMI accident has been carried out during refuelling shutdowns which appear to have increased in length somewhat since 1979.

2-loop units. The performance of 2-loop units has, in every respect, been vastly superior to that of larger units. This is reflected not only in good unit maturation, leading to sustained high capacity factors, but also in low shutdown frequencies and the absence of operation errors and regulatory limitations. Although all the units have suffered severe steam generator corrosion, including the need to replace the steam generators at Point Beach 1 in 1984, and a serious tube leak in the Ginna unit in 1982, the utilities have succeeded in restricting many repairs to the annual refuelling shutdown and completing repairs quickly – the steam generator replacement at Point Beach was carried out in about three months rather than the more typical period of about a year.

3-loop units. In overall capacity factor terms, the performance of the 3-loop units is little different to that of the 4-loop units. However, closer examination of the data shows that this group has suffered far more seriously from steam generator problems. All five units that were declared commercial before 1974 suffered chronic steam generator leakage problems resulting finally in the need, in all cases, for steam generator replacement. These units seem to have suffered not only from the generic problem but also the change in water treatment from phosphate treatment prior to 1974 to AVT. Given that these problems are likely to have been more a result of the time of start-up of these units than the number of coolant-loops, it may be that the capacity factors for 3-loop units have been lower than would be representative of this size group. Regulatory limitations were almost entirely incurred in 1979 and resulted from shutdowns of five months and seven months required to verify the adequacy of piping supports at two units.

4-loop units. The 4-loop units remain the most difficult group to analyse, their main feature being the inconsistency of experience apart from the usual similarity of performance of twin reactors on the same site. The eight units with a substantial length of service behind them seem to fall into three groups. The first group (comprising two twin units) have performed consistently and at well above average capacity factors – note nevertheless that the performance of even this pair falls short of the performance of the *worst* 2-loop unit. The second group also containing two twin units has performed less consistently and at lower mean capacity factors. The predominant feature of the four units in the third group is the extreme variability of performance and, for the worst of them, very low mean capacity factors.

Unlike 3-loop units, where a great deal of the poor performance and variability can be traced back to specific causes, most notably steam generator problems, the causes of poor performance in the 4-loop units are much more diverse. The main problems have been:
– steam generator corrosion. This has not been as serious as with 3-loop units, perhaps because the importance of careful control of water chemistry was beginning to be appreciated when most of the 4-loop units entered

service, but has nevertheless caused significant problems. Severe pitting at Indian Point 3 and stress corrosion cracking at Cook 2 may lead to the replacement of the steam generators at both units in 1989
- corrosion of condenser tubes due to salt water cooling. This has been a factor at three of the 'problem' units of the third group
- turbine generators. As was originally expected, the large slow-speed wet-steam turbine generators required by this size of unit have proved technologically demanding. Unlike the smaller units where breakdowns have been concentrated in the early years, for units of greater than 1000 MW breakdowns have persisted. Note that this problem is not so much a function of the NSSS supplier (which does not necessarily supply the turbine generator) as a function of the output size and make of the turbine generator
- construction and installation errors. A number of shutdowns (usually classified as regulatory limitations) have been necessary especially at the Trojan plant to rectify construction and installation errors and to verify the integrity of plant systems

Overall the question of whether there is a 'size effect' remains tantalisingly difficult to establish. In engineering terms, this could be rationalised in a number of ways. If a major determinant of performance was failures in the coolant loop then, simply by virtue of there being more coolant loops, it would be expected that large units would be less reliable than small units. Alternatively, if a major determinant of performance was the reliability of the 'non-modular' components such as the turbine generator, then large components may be intrinsically more difficult to fabricate (and potentially less reliable) than smaller components. Either of these hypotheses would tend to lead to a decline in performance with increasing size so that 2-loop units should outperform 3-loop units which in turn should outperform 4-loop units.

A somewhat less precise reason that has been put forward to explain the size effect is complexity. This theory suggests that larger units perform less well due to their complexity. This might operate in a number of ways. At a managerial level, large units will require larger workforces to construct, operate and maintain and present more difficult problems of communication and management. At a technical level, larger units may require more control systems all of which increase the potential for equipment failure.

As well as these factors, geographical and temporal factors may also have been important. All six 2-loop units are located in areas identified earlier as containing the best-performing units – five are in the Wisconsin–Minnesota region and one in New England. The utilities, although relatively small, have developed very high reputations for engineering excellence and a number of individuals who were at the centre of the utility management team in the construction phase now hold senior positions in national bodies such as EPRI and INPO. A good example of the foresight of these utilities was the decision in the design of the Point Beach units to allow sufficient space in the containment of the steam generators to be readily replaced. This allowed them to be replaced in a little over three months rather than the year which most of the replacements have required. This extra space is also likely to have allowed easier access for maintenance and thus reduced maintenance times. The apparent competence of these utilities may also have allowed the NRC to treat them more flexibly and impose fewer regulatory restrictions.

The temporal factors which may account for the size effect relate to the early completion of the 2-loop units which were all declared commercial in the period 1970–74 before many of the 3-loop and particularly the 4-loop units were completed. It has been suggested that the earlier units were not subject to as many design changes and fitting of additional safety systems during their construction phase as units completed from the mid-1970s onwards – they are sometimes spoken of as the 'race-horses' of the reactor world. The design changes and additional systems may have added to the complexity of units bringing about unexpected interactions. Additionally, these additional systems may have filled up the containment and made access more difficult and hence maintenance slower. In all but the orders from 1973, the containment was deliberately reduced in size as a cost-saving measure. Whilst these temporal factors are intuitively plausible, it should be noted that the first 3-loop units were almost exactly contemporary with the 2-loop units but even before they all required steam generator replacement, their operating record was markedly inferior.

No single theory seems sufficient to account satisfactorily for the remarkable and consistent size effect by itself but on the other hand none can be discounted as insignificant. This effect is of some importance in technology choice for the future and we return to it later.

General Electric units

Two outstanding features emerge from the performance of GE BWRs. The first is the variability of each unit from year to year; characteristically, a unit achieving a high capacity factor one year will achieve a very low capacity factor in the following year. Thus, few BWRs have very high life-time capacity factors and, equally, few have very low life-time capacity factors – perhaps predictably the two units with the best record are situated in Wisconsin and New England.

The second feature is the severe decline in capacity factors in the 1980s. This has resulted largely from the need to carry out lengthy repairs to

Table 5.4. *Performance of mature General Electric units by time period*

Period	500–700 MW		700–950 MW		950+ MW	
	Capacity factor (%)	No. of reactor years	Capacity factor (%)	No. of reactor years	Capacity factor (%)	No. of reactor years
1976–78	67.2	18	63.3	18	59.7	7
1979–81	64.5	21	56.3	26	67.1	14
1982–85	57.6	28	54.7	40	44.1	20
1986	53.1	7	59.5	10	31.4	6[a]

[a] Due to a serious lack of confidence in the TVA's nuclear plant management, NRC and TVA agreed in August 1985 that TVA's nuclear plant should be shut down. No firm date for the plant's restart has yet been set. The five affected units include three large GE BWRs (Browns Ferry 1–3) which had in fact been shut down since March of that year. If these units are discounted, the average for mature large GE BWRs in 1986 rises to 62.8%.
Source: SPRU nuclear power plant data base.

combat cracks in the pipe-work resulting from the phenomenon known as intergranular stress corrosion cracking (IGSCC) (see below).

The performance of US BWRs is summarised in Table 5.4 and Table 5.5 and shown in greater detail in the Appendix to this chapter. These tables analyse performance in the same way as those relating to Westinghouse units.

Intergranular stress corrosion cracking. In many ways, the IGSCC problem parallels the PWR steam generator problem, involving a lengthy period of research to fully understand the problem, multi-facetted remedies and long shutdowns to repair severely affected units.

The first indication that pipe cracking was a problem for BWRs came in 1974 with the discovery, in a US unit, of cracking in the main heat removal system. Subsequently cracks have been found in a number of plant piping systems with cracking in the large diameter recirculation pipes causing particular concern in recent years. Three factors are now thought to be necessary precursors to IGSCC:

– weld sensitisation. High carbon content in steel makes it more susceptible to IGSCC and, although low carbon steels were used in BWR pipe-work, the welding process increased the carbon content at the point of the weld
– high tensile stresses. During installation, additional stresses may be placed on the piping which increase its susceptibility to cracking
– water chemistry. Inevitably BWR coolant water contains a high level of dissolved oxygen which produces a very hostile environment. The presence of other impurities, particularly sulphates, exacerbates the problem

Table 5.5. *Performance of General Electric units by unit age*

Age	500–700 MW		700–950 MW		950+ MW	
	Capacity factor (%)	No. of reactor years	Capacity factor (%)	No. of reactor years	Capacity factor (%)	No. of reactor years
1	55.5	7	56.1	10	46.0	11
2	53.5	7	50.2	10	50.8	9
3 onwards	61.7	88	56.9	99	51.6	47

Source: SPRU nuclear power plant data base.

For operating plants suffering from this problem, the utility must either replace the piping with a more corrosion-resistant material – usually nuclear grade stainless steels with low carbon content – or repair the weld using an 'overlay' usually equivalent in thickness to the original pipe wall. More generally, utilities must be aware of the importance of good practice in welding and of good water chemistry, particularly the avoidance of impurities. Controlling the level of dissolved oxygen by injection of hydrogen is increasingly being adopted.

For plants with welds which are not yet cracked, but which are at risk, a process known as induction heating stress improvement has been developed. This process was designed to reverse the effects of tensile stress incurred during installation and may be a useful preventive measure even on units with nuclear grade steel piping, although there are now doubts about its effectiveness.

The appreciation of how widespread the problem is, and the establishment of methods to combat and avoid it, have been relatively recent and the detailed shutdown analysis (which is based on data to the end of 1984) does not fully reflect the impact of the problem hinted at in the very low capacity factors of the most recent years. Current expectations are that almost all plant installed more than five years ago will have to (or has already had to) undergo major pipe replacement at a direct cost of about $50m per plant in addition to the larger replacement power costs resulting from a shutdown of six months or more. Weld overlays may continue to be used mostly as a short term holding measure..

Decontamination. One of the side effects of the IGSCC work has been to throw greater emphasis on decontamination methods. Since BWRs have a direct steam cycle (i.e. the core coolant, which picks up a small amount of radioactivity, is used to drive the turbines) radioactive contamination is much more wide-spread than in PWRs which have a secondary steam circuit. For complex, labour-intensive repairs such as pipe-work replacement, decontamination is often necessary to minimise dosage rates to the repair crew. However, these techniques are still evolving and there are fears that some current techniques may have an adverse long-term effect on some materials.

Unit maturation. The onset of IGSCC after five or more years tends to mask the improvement in performance that occurs over the first two to three years. In particular shutdown frequency and operating losses tend to fall, as do shutdowns due to failures in the turbine generator and reactor control system and for testing. Factors such as the IGSCC problem mean that, unlike PWRs, failures in the nuclear auxiliary and emergency systems, main heat removal system and feedwater condenser and circulating water systems are persistent and do not decline as the unit gets older. There have been few shutdowns due to problems in the reactor and accessories, the fuel and for training and licensing.

Regulatory limitations. Generally the lengthier regulatory limitations have been necessary to carry out inspections and testing to establish whether IGSCC has occurred. In addition some shutdowns have been required to check the adequacy of pipe restraints.

The Browns Ferry fire. The Browns Ferry fire, discussed in the previous chapter, does distort the performance data for 950+ MW units somewhat. The fire resulted in a shutdown of about 18 months in Browns Ferry 1 and 2 (classified as a shutdown due to operation error). This shutdown is reflected in the performance in years 1 and 2 of unit 1 and year 1 of unit 2, and also in the performance in 1975 and 1976.

   Overall, it is difficult to establish a clear picture as to the determinants of BWR performance. There seems to be little evidence for a size effect and, even before the impact of IGSCC on performance became overwhelming, the variability of performance from year to year at the same unit was high with little pattern to the equipment problems emerging.

Babcock & Wilcox units

In statistical terms, the performance data on Babcock & Wilcox units is very heavily conditioned by the TMI accident. For economic assessment purposes it would seem appropriate to include these units. However, the objective of this analysis is to elucidate the determinants of performance, including trends on maturation and such trends will only be obscured if

Table 5.6. *Performance of mature Babcock & Wilcox units by time period*

Period	Inc. Three Mile Island		Exc. Three Mile Island	
	Capacity factor (%)	No. of reactor years	Capacity factor (%)	No. of reactor years
1976–78	64.9	12	64.9	12
1979–81	46.9	22	56.7	18
1982–85	45.6	36	58.3	28
1986	41.1	9	43.0	7

Source: SPRU nuclear power plant data base.

data from units which were shutdown for the whole period are included.*
The remaining seven units are still not a homogeneous sample with over
half of the operating performance having been accumulated at the Oconee
site where three essentially identical units are installed. These units are
operated by Duke Power, which has a good reputation for engineering and
managerial competence and, particularly in recent years, they have performed well – the mean capacity factor for the three Oconee units in 1984
exceeded 80%. By comparison the record of the other four units is much less
impressive.

The operating record of the B&W units is summarised in Table 5.6 and
Table 5.7 and shown in greater detail in the Appendix to this chapter. These
tables analyse performance in the same way as those relating to Westinghouse units. The main features are:
- the overall level of performance, if the TMI units are excluded, is no
  worse than average with reasonable evidence of unit maturation
- the TMI accident appears to have had a severe impact on the performance of the other B&W units with a trough in performance being reached
  in 1982 and performance climbing back to pre-accident levels by 1984
- closer inspection of the precise causes of this trough show that, whereas
  in 1979 (the year of the TMI accident) there were lengthy regulatory
  shutdowns, in the period 1980–82, the primary cause of poor performance
  was lengthy maintenance and repair shutdowns. These may in part have
  been due to the coincidence of 10-year inspections at the Oconee units
  and also a requirement to carry out a large number of backfits during the
  maintenance and refuelling shutdowns

* TMI 2 was declared commercial in December 1978 only four months before the accident,
  and it is excluded entirely from the detailed analyses. TMI 1 was not involved in the accident
  but was immediately shutdown and did not re-enter service until Autumn 1985. This unit
  was declared commercial in September 1974 and its performance data for 1975–78 are left
  in, on the grounds that these data are representative of B&W units, but the data from 1979
  onwards are omitted.

Table 5.7. *Performance of Babcock & Wilcox units by unit age*

Age	Inc. Three Mile Island		Exc. Three Mile Island	
	Capacity factor (%)	No. of reactor years	Capacity factor (%)	No. of reactor years
1	49.1	9	53.2	8
2	49.6	9	55.8	8
3 onwards	48.8	79	57.4	65

Source: SPRU nuclear power plant data base.

– the steam generators have been an increasingly significant cause of shutdowns due to cracking near the upper ends of the tubes. B&W steam generators are of significantly different design to those of Westinghouse or Combustion Engineering being a 'once-through' rather than U-tube design. The use of AVT from commissioning and a stress relief treatment after assembly has meant that B&W units have not suffered significantly from denting, wastage or stress corrosion cracking. The reasons for the cracking that has occurred are not entirely clear and although B&W has recommended measures to combat it, it is too early to judge whether they will be effective
– equipment failures in the main heat removal system and the turbine generator seem persistent, whilst failures in the reactor control system appear to decline with age
– failures in the reactor and accessories, although higher than with other vendors' units, are much lower in recent years, with most of the shutdown time attributable to a particular problem at the Oconee units in 1976

Overall, B&W units have tended to be unreliable with high shutdown frequencies and a number of persistent problems, but recent experience suggests that good performance is possible from these units if the utility is prepared to devote sufficient resources to their operation and maintenance.

There has been a disproportionately high incidence of events with potentially serious consequences at B&W units. Apart from the TMI accident, there have been two serious incidents at the Davis Besse plant, one involving many of the conditions which led to the TMI accident. However, the number of units and incidents is too small to determine whether that is a statistical anomaly or of real significance.

Combustion Engineering units

Overall, the Combustion Engineering units have the best operating record of the four US vendors. A particular problem with analysing their perform-

Table 5.8. *Performance of mature Combustion Engineering units by time period*

	Inc. Palisades		Exc. Palisades	
Period	Capacity factor (%)	No. of reactor years	Capacity factor (%)	No. of reactor years
1976–78	63.7	10	69.6	7
1979–81	64.4	20	68.1	17
1982–85	65.2	31	67.8	27
1986	73.9	10	80.8	9

Source: SPRU nuclear power plant data base.

ance concerns the treatment of the data relating to the Palisades unit, the first unit of this make to enter service, and which has performed consistently poorly. It has been decided, on balance, to omit this unit from the detailed analyses for a number of reasons:
- its design differs in a number of important respects, such as the control rod drive mechanism, from its successors
- perhaps reflecting these design changes, the performance pattern differs markedly from its successors. In particular it has suffered far more problems with the steam generators and the reactor control system, and it has never been able to operate at its design rating – currently it is operating at less than 80% of design rating
- the first unit(s) built by other vendors which were ordered prior to Oyster Creek have been omitted from the analysis as not being representative of current designs. It is likely that these units incorporated the sort of first-of-a-kind errors that have occurred at Palisades and thus to include the Palisades unit would be to penalise Combustion Engineering unfairly

Apart from the Palisades unit, the variability of performance is low with the long-term capacity factors of all units beating the industry mean. The operating record is summarised in Table 5.8 and Table 5.9 and shown in greater detail in the Appendix to this chapter. The main features are:
- there is good evidence of unit maturation with shutdown frequency, operating losses and the reliability of a number of systems such as the feedwater, condenser and circulating water system improving with age
- overall capacity factors have been little affected by the TMI accident
- steam generators have proved reliable with few of the shutdowns attributed to equipment failure in the steam generator being due to leakage. All the Combustion Engineering units except Palisades have used AVT from their commissioning and the only major problem has been 'pitting' at one unit which required sleeving during a refuelling shutdown

Generally Combustion Engineering reactors seem to have fewer technical question marks against them than those of the other vendors and, with

Table 5.9. *Performance of Combustion Engineering units by age*

Age	Inc. Palisades		Exc. Palisades	
	Capacity factor (%)	No. of reactor years	Capacity factor (%)	No. of reactor years
1	58.0	11	61.2	10
2	60.1	11	62.7	10
3 onwards	64.6	74	69.9	61

Source: SPRU nuclear power plant data base.

consistently good operating performance as well as good construction performance at the recently completed St Lucie and Palo Verde units, their reputation is currently high.

Capital costs and construction times

The data on construction times and costs for the USA are much more plentiful than for the other three countries studied in this book. However, there are two important limitations on the weight which can be placed on this data. The first is technical; the data, while abundant, are often suspect and not always in an analytically useful form. It has proved impossible, for instance, to obtain a comprehensive series of dates for the start of construction activity at US nuclear sites, and capital cost data are even more suspect. Secondly, the first era of nuclear power construction has now largely ended in the USA; if there is to be a second era, it will have limited historical continuity with the first, and therefore the lessons to be learned will be correspondingly limited.

It is also important to bear in mind that the USA is, as emphasised in the previous chapter, a large and diverse country. Some 60 utilities own (or part own) nuclear plants, and regional variations, in such matters as labour costs or unionisation rates, are much wider than in the other countries studied. This diversity means that achieving fair comparisons in construction experience between, say, New England, Florida and California is, in some respects, nearly as difficult as between different countries.

Despite some shortcomings, the data on construction times are, as for the other countries, better than on construction costs. Table 5.10 shows construction times by year of construction start. This series includes all commercial-sized plants (over 450 MW) ordered since (and including) Oyster Creek in 1963. This means that it includes all the 12 early 'turnkey' plants which were built for a fixed price. It is appropriate to include these

Table 5.10. *US average construction times by year of construction permit issue*

Year of construction permit issue	Average length of construction time[a] (months)	No. of units
1964	53	2
1965	–	–
1966	66	5
1967	69	14
1968	84	21
1969	82	8
1970	98	9
1971	89	4
1972	86	3
1973	124	9[b]
1974	148	9[c]
1975	132	5[d]
1976	125	5[e]
1977	100	4[f]
1978	(116)	(1)[g]

[a] Defined as time between issue of construction permit and point of commercial operation (handover to utility).
[b] *Excluding* three plants, Shoreham and Watts Bar 1 and 2, which are indefinitely stalled.
[c] *Including* five plants which are not yet commercial but which are expected to become commercial by early 1988: *excluding* four plants – Bellefonte 1 and 2 and Comanche Peak 1 and 2 – which are indefinitely stalled.
[d] *Including* two plants which are not yet commercial but which are expected to become commercial by early 1988: *excluding* three plants – Braidwood 1 and 2 and South Texas 2 – which are indefinitely stalled.
[e] *Including* two plants which are not yet commercial but which are expected to become commercial by early 1988: *excluding* one plant, Seabrook 1, which is indefinitely stalled.
[f] *Including* one plant which is not yet commercial but which is expected to become commercial by mid-1987.
[g] The only plant given a permit in 1978 was Harris and it is expected to become commercial in late 1987. No plants have been given construction permits since 1978.
Source: IAEA (1985) *Nuclear power reactors in the world*, April 1985 Edition, Vienna; Electric Power Research Institute (1984) *An analysis of power plant construction lead times*, Volume 2: Supporting documentation and appendixes, EPRI EA-2880, Palo Alto, February 1984; *Nucleonics Week* various 1986 and 1987 issues.

plants, because while their declared cost figures are distorted by the fixed nature of the contracts (on which the vendors reportedly (United States Atomic Energy Commission, 1974, p. 27, p. 30) lost large sums of money), their construction schedules were subject to the same forces as other plants of the early commercial period.

Because of the diversity of sizes, manufacturers, architect-engineers and locations, the data in Table 5.10 are of limited value without 'normalisation' to allow for some of these factors. The importance of Table 5.10 is that it attempts – unlike other analyses of US construction times – to put US experience on a basis identical to that used for the other countries in this book: namely, to express construction times from the start of serious construction on site (first concrete) up to the point of commercial operation. Most other studies of the USA report on a different basis, either using construction permit application as the start point, and/or fuel loading or first power as the completion point. From a utility's (economic) point of view, the period chosen here is the most appropriate as it represents the time over which significant capital is tied up without return.

The problem encountered in compiling Table 5.10 is that there is no reliable series which gives a full set of start of construction dates. The excellent EPRI study of 1983 (Electric Power Research Institute, 1984) gives 'first concrete pour' for some but not all plants; the main IAEA source (International Atomic Energy Agency, annual) gives a number of dates for construction start which are clearly inconsistent with detailed case study material in the EPRI study and this is also true for the construction start dates given by US Department of Energy sources (which may in fact refer to initial site work; this can precede first concrete by months or years).

As a consequence of these difficulties, the data in Table 5.10 use date of issue of construction permit as a proxy for date of serious start on site. Although issuing of construction permit may occur several years after application it is not likely to occur until the utility is ready to proceed with construction. The problem is not only that this does not exactly correspond to start date, but that the bias to which it gives rise varies through time. Prior to the 'Calvert Cliffs' regulatory decision of 1971 (see previous chapter) utilities often began site work before a permit was issued. After 'Calvert Cliffs' came into operation, with its requirement that an Environmental Impact Statement be produced for each proposed unit, construction could not begin until the construction permit was issued, and was frequently delayed for several further months.

These biases mean that the times reported in Table 5.10 for the earlier reactors (up to 1972 permits) in general understate the true construction period by a few months, while the times for reactors of 1973 vintage and later are similarly overstated. This means that while the main trend in lead times – steadily upwards in the decade 1964 to 1974 – is confirmed, its steepness is somewhat less than Table 5.10 seems to show. In addition, the sharp break between 1972 and 1973 is more a consequence of the implementation of the Calvert Cliffs decision than of a fundamental underlying shift at that time.

With all these qualifications entered, it remains the case that construction times roughly doubled in the decade from the mid-1960s, and have – for those reactors not deliberately delayed – since that time begun to reduce somewhat. A large number of studies (among them Mooz, 1981, pp. 197–225; Komanoff, 1981; Electric Power Research Institute, 1984), many of them using multiple regression techniques, have sought to explain these trends in terms of such variables as size of plant, location, presence of cooling towers, duplicate or twin units and experience of utility or architect-engineer. These variables have been found to have some explanatory power, but all the studies have been left with a large residual time trend, at least for reactors started up to 1974, which it has not been possible to capture using any plausibly explanatory variable in the regressions.

There has been widespread agreement that the fundamental time-related force has been regulatory. In the earlier, 'Komanoff' hypothesis (1981), increasing stringency of regulation is related to the existing size of the nuclear sector. Regulators wish, according to this view, to ensure that the *total* risk of reactor operation is kept constant, but this implies stricter regulation of each *individual* reactor as the numbers increase. This increased stringency directly leads to longer construction times (and higher costs). If this hypothesis were correct construction times should get longer for each successive year of reactor start. In practice, however, more recent reactors have been constructed rather more quickly than those of 1974/75.

This fact lends support to the more recent EPRI regulatory explanation, which is that 'normal' regulation has been getting marginally more stringent with time (a phenomenon with which constructors did, after initial difficulty, cope reasonably easily) but that, superimposed on this normal pattern, there have been three regulatory 'shocks' in 1971, 1975 and 1979.* These shocks, particularly the first, had their most damaging influence on plants designed in the late 1960s and early 1970s whose construction permits were delayed until 1973 and 1974 and which then had to be successively re-designed to cope with the 1971 and 1975 shocks. Plants licensed in the later 1970s were able to absorb most of those lessons, and, while the repercussions of TMI extended their schedules somewhat, their

---

* Electric Power Research Institute, 1984. Vol 1, pp. 2–3 to 2–7. These 'shocks' were the Calvert Cliffs decision of 1971, regulatory reactions to the Browns Ferry fire of 1975, and the TMI accident of 1979.

Table 5.11. *US average construction times by reactor size and vendor (months)*

Reactor vendor	Reactor size		
	Small[a]	Medium-sized[b]	Large[c]
1. Reactors with construction permits up to 1972			
Westinghouse	56 (7)[d]	81 (11)	119 (11)
General Electric	56 (6)	69 (9)	89 (5)
Babcock & Wilcox	—	84 (9)	—
Combustion Engineering	72 (1)	82 (7)	—
Average	57 (4)	79 (36)	110 (16)
2. Reactors with construction permits in 1973 or later			
Westinghouse	—	—	122 (7)
General Electric	—	—	130 (6)
Combustion Engineering	—	75 (1)	121 (4)
Average	—	75 (1)	125 (17)

[a] Small reactors are, in the case of Westinghouse, 2-loop machines and, for other vendors, units under 700 MW.
[b] Medium-sized reactors are, for Westinghouse, 3-loop machines, and for other vendors, in the 700–950 MW range.
[c] Large reactors are, for Westinghouse, 4-loop machines, and for other vendors, over 950 MW.
[d] Numbers in brackets refer to number of reactors in the group.
Source: As for Table 5.8.

adaptation to the more settled regulatory climate of the early 1980s has allowed them to be built more rapidly than 1974/75 units.

This EPRI explanation of the time trend seems as convincing as any available, and it does not ignore the other important time-related trend in construction times – the tendency for utilities to 'stretch out' construction times deliberately, because of financial difficulties, or low load-growth, or both. In practice, these stretchouts have not applied significantly to the majority of units. Only five of the 33 post–1972 units shown in Table 5.8 had experienced voluntary stretchouts of longer than six months by 1983, an identical percentage (15%) to the earlier plants experiencing significant voluntary stretchout (Electric Power Research Institute, 1984, Vol. 1, pp. 2–17 to 2–19). The main brunt of stretchouts has been borne by the 11 post–1972 plants omitted from Table 5.10, and by a large number of other units which have, by now, been cancelled altogether.

Among the other factors influencing construction times, those found most significant in other studies have been size of units, twin units and constructor experience. A universal finding has been that size of unit is a highly significant determinant of construction time, with smallest units taking much the shortest time to build. Table 5.11 shows the data on this subject, organised by manufacturer of NSSS and plant sizes that are consistent with those used in the operating performance section. This shows a very clear trend towards longer length of construction with increased size of unit: the early small reactors were built particularly rapidly, and this partly accounts for the frustration within the American industry as times lengthened for larger (and later) units. There is no indication that NSSS manufacturer has any significant effect on construction times. Among early General Electric plants, the medium- and large-sized plants were built 12–20 months faster than their competitors' plants, but no statistically significant effect emerges in the various regressions and the effect is in any case absent from later plants.

Twin units also have a significant impact on construction times. Among the 23 pairs of genuinely twin units (as opposed simply to common siting) there were 21 cases of second units taking longer to construct. The range was wide – from 2 to 52 months' excess time – and the average excess among second units was fairly high, at 17 months. It is however important not to over-interpret this finding. For those second units started simultaneously with first units (20 out of the 23), a large part of the explanation will be in terms of deliberate scheduling policy. The chapter on France makes it clear that EdF put their first unit on the critical path and aim to take longer on second units as a cost-minimising strategy. This will almost certainly have been the case for a number of the US units. Often the intention will have been to complete second units closer in time to first units than the average of 17 months, but in many cases the size of the gap may be due partly to the fact that later units tend to take longer to build than earlier ones. But in the end, the statistical problem should lead to the major caution about these data. While 20 out of 23 twins received *construction permits* for each unit at the same moment (issuing of construction permits being the proxy used in this chapter to describe construction start) it is by no means certain that serious site work will necessarily also have started simultaneously. The true excess time for second units may well be substantially less than the apparent average.

The final factor affecting construction times is learning. This has been investigated in a number of studies (Mooz, 1981; Komanoff, 1981; Zimmerman, 1981), all of which found that learning effects could be identified only through the experience of the architect–engineer. The learning effect has been modest – for each doubling in the number of plants a fall of 7 to 8% in construction time (Mooz, 1981, p. 209). This is not surprising given the diversity and 'customising' of much US nuclear design, nor is it likely to be of much future relevance as any future nuclear plants are likely to differ significantly from those of the past and there will have been a large time gap.

Turning to construction costs, there is a vast amount of capital cost data for US reactors, most of it entirely valueless. This is the 'soft' data on expectations about future capital costs of plants, often notional, about to be built.* These data might have some limited use if they had proved a reasonable guide to actual costs. However, the forecasting record has been extremely poor – large and consistent underestimates of actual costs – and, furthermore, there has been no evidence of forecasters learning from early underestimates. The estimates were of a narrowly 'engineering economics' type and while they successively incorporated past cost increases they always and erroneously assumed that there would be no *future* increases: they ignored trends. Consequently this large body of data should be ignored.

The more limited quantity of 'hard' data or actual, historical capital costs itself also has serious problems. There is a public source in the annual returns that utilities must make to the US Department of Energy (now reported in Energy Information Administration, *Historical plant cost and annual production expenses for selected electric plants* (annual), Washington DC). However, the questionnaire forms have changed over the years, reporting was not originally mandatory, the data collected are highly aggregative and the notes accompanying the questionnaire have been insufficiently detailed to ensure that there will be a common reporting procedure. To this is added the problem that utilities report in 'mixed current dollars' (i.e. at the price levels prevailing in each year, and incorporating a range of non-explicit interest rates). Among the more detailed difficulties is that for twin units – which make up the majority of plants completed – it is quite unclear what accounting conventions have been used by different utilities to split costs between units.

The response of the main students of capital cost, Komanoff, Mooz and EPRI, has been to make some broad and occasionally heroic attempts to convert the data into constant dollar terms, remove interest during construction and conduct limited but detailed cost surveys with co-operative utilities (most of the results of which have remained private). These efforts have yielded some very broad trends, though it would be extremely dangerous to try to go beyond them.

---

* The best known of these 'soft' data is found in a long series of architect-engineer studies of costs of a notional nuclear plant for future construction. This series was originally sponsored by the United States Atomic Energy Commission: the first publication was 'WASH-1082' *Current Status and Future Technical and Economic Potential of the Light Water Reactor*, USAEC, 1968 (March). This series continued through WASH-1150 (1970), WASH-1230 (1972) and WASH-1345 (1974). The architect-engineers United Engineers and Constructors have continued this work in ever more elaborate forms: a more recent example is found in US National Technical Information Service, *Phase IV Update (1983) Report for the Energy Economic Data Base Program*, US Dept of Commerce (updated).

Table 5.12. *Construction costs for completed US commercial reactors (excluding interest) by year of construction permit issue (1982 $m/kW)*[a]

Construction permit issue date	Cost	No. of plants
1967	488	9
1968	782	21
1969	925	7
1970	959	9
1971	845	4
1972	1 107	3
1973	(1 358)[b]	9

[a]This table excludes the twelve early 'turnkeys': because they were contracted at fixed cost, and the vendors made large losses, the published cost figures are a poor guide to real resource use.
[b]The 1973 figures include estimated rather than actual cost for seven of the nine plants. Historically this will more likely lead to understatement than overstatement of actual final cost. No completed plant data were available for 1974 starts (or later).
Source: EPRI (see Table 5.10) Vol 2 pp. D–13 to D–15.

Table 5.12 shows the data for construction cost (at 1982 prices without interest, as corrected by EPRI) by year of construction permit issue for plants completed to 1983. As in the case of construction times, there is clearly a powerful time trend, such that real costs per kW rose at a rate of over 170% in only six years, or some 18% per year in *real* terms.

The most painstaking and careful work in capital costs is undoubtedly that of Komanoff (1981). Although he finds no direct statistical link between construction times and costs he explains the time trend for costs in terms similar to those used for construction times: regulatory pressure aimed at stabilising total risk by reducing individual plant risk. Komanoff assumed that the upward trend in cost would continue unabated. The evidence from more recent plants is not clear: costs have not fallen in the way that construction times have but the rate of increase appears to have slowed somewhat (EPRI, 1984).

Komanoff attempts to explain other non-time related causes of variation in capital cost. Among his findings are that twin units collectively cost less than single units, that higher unit sizes lead to very small scale economies, that architect-engineer experience had a modest learning effect and that north-east location was associated with much higher costs per kilowatt (for non-nuclear, labour market reasons). These various relatively small effects were swamped by the dominant time trend and, given the weakness of the raw data, not much weight should be placed on them.

Overall then, US cost and time experience exhibits a very clear and strong upward trend for plants licensed to be built between the mid-1960s and the early 1970s. For those relatively few plants licenced from 1975 to 1978 (the most recent licensing date) construction times have begun to fall again, but there is no evidence that costs have stopped rising though the rate of increase has almost certainly slowed. This experience of the first nuclear era – now, in construction terms, drawing to a close – is of limited relevance to any future US nuclear programme, but it has significantly contributed to the difficulties of envisaging a second nuclear era. Decision-makers have become sceptical about the open-ended and often uncontrollable economics of nuclear power and official US Department of Energy figures have shown the economic balance between coal-fired and nuclear plants shifting heavily in favour of coal-firing in the 1980s. By 1986 there was no region left in the USA where (even supposing a new nuclear plant could be built) it would be more economic to build a nuclear than a coal-fired plant (see Nuclear Energy Agency, 1986), and in the regions of cheapest coal, the costs of nuclear electricity were expected to approach double the costs of coal-firing.

INSTITUTIONAL INFLUENCE

Compared to its international competitors and to expectations, the US nuclear programme has not been a success judged by the key criteria of operating performance and construction costs and times – although by other criteria such as oil substitution it may be judged more successful. More than 20 years after the first commercial order was placed, the technology cannot be counted fully mature – it is not clear that serious generic faults have been solved, designs are not stable and construction costs and times are uncertain. For the future, with no orders placed for over a decade that will actually be completed, there is now the prospect that the industry will soon be moving down a 'forgetting' curve rather than up a learning curve. This may already be happening in the design and component manufacture sectors, it seems imminent in the site construction and assembly sectors, with only the plant operation and maintenance sector cushioned against this danger by the large stock of complete or nearly complete units.

None of the institutions involved in nuclear power can escape some element of criticism. In addition, the political environment, for example the strong opposition movement and the dominant influence of market forces, may not have been conducive to the efficient development of nuclear power. The influence of public opposition should not however be exaggerated and used as an excuse for the failings of the nuclear industry. Even in the post-TMI environment, a number of US utilities have demonstrated that nuclear plant can be built to time and cost and operated at high capacity factors, so whilst the environment may be difficult it is by no means impossible.

Prior to dealing with the institutions in turn, it is worth going back to the commercialisation phase of 1963–74 since much of what has happened since then has been conditioned by events in that period, particularly its early part. The flood of orders and the competitive pressures it brought with it had a number of important effects:

– for all but the latest orders, there has been no chance to profit from the important feedback loop by which 'learning by experience' is incorporated in subsequent design, construction and operation

– the flood of orders placed very heavy demands for skilled personnel which it may not have been possible to meet satisfactorily. Manufacturers may have been forced to devote their skilled personnel to carrying through existing orders rather than developing second and third generation designs which should have overcome the generic faults inherent in the first generation

– the proliferation of orders encouraged the belief that nuclear power was a fully mature technology and could be handled in much the same way as fossil-fired plant. This resulted in a number of utilities and architect-engineers entering the field with an inadequate appreciation of the task they were undertaking. Practices which had proved viable in the construction and operation of fossil-fired plant, such as filling out the design details after the start of construction and neglecting factors such as water chemistry and metallurgy, in the knowledge that, for example, a leaking boiler tube could be readily replaced, were not appropriate to a nuclear plant due to the complexity of designs and the difficulty of access to much of the plant

– competitive pressures amongst the vendors led them to adopt measures to reduce unit capital costs which now seem ill-advised. In particular the very rapid scaling up of unit sizes and the reduction in containment volumes seem to have had adverse effects

– the task of regulating the design and construction of these orders was gigantic. Whether the AEC did not appreciate the scale of the task or did not have the political capability to stand out against the bandwagon of nuclear ordering is not clear. Nevertheless the effect has been that basic design issues that should have been resolved in the first handful of orders are only now being resolved. At individual plant level, issues in detailed plant design which should have been addressed before construction work started have arisen during construction often causing severe disruption to schedules. Such issues have often led to backfitting requirements which are sometimes wrongly blamed on regulatory action

– there has been a vicious circle effect with lengthy lead-times and design changes. As construction times lengthened, in the mid-1970s often due to utilities' financial difficulties each unit became subject to increasing design changes and additional regulatory requirements which in turn lengthened lead-times and increased costs. Whilst in isolation these changes may have made good sense, there are fears that the effect on the morale of the workforce and hence the productivity and quality of its work may have reduced the net benefits

Many of these problems – generic faults in the first generation of a given technology, the weeding out of firms not equipped to handle the technology etc. – are not unusual and may be inevitable in the commercialisation process of a complex, demanding technology. However, the pace of ordering has left the USA with a stock of reactors on which the technology will inevitably be judged and which, ideally should have encompassed several generations of design and learning, but which is in fact drawn from not much more than the first design generation.

Turning to the individual institutions, the failings of the utilities can be summarised as not being informed, critical buyers and of not appreciating the interdependence of utilities and the benefits of their acting together. It has often been suggested that the successful utilities would be the large ones with strong in-house engineering capabilities but whilst this appears to be borne out in other countries (e.g. Ontario Hydro in Canada), US experience suggests that size is not ultimately the determining factor. Many of the best units are operated by relatively small utilities such as Wisconsin Electric Power, whilst the record of the largest utilities, TVA and Commonwealth Edison, has been no better than average. An explanation may be that the requirement is to place people of high quality in the key management positions. These people can then select and manage the design and construction teams, be they bought-in through A-Es and constructors, or from within utility resources, as Duke Power is doing with increasing success.

The importance of utilities acting together is two-fold. First, if relatively small utilities are to exert significant market power against vendors, A-Es etc., it is important that learning be shared and that utilities pool resources to carry out the relevant R&D. This is now being increasingly recognised and is evident in the utilities' decision to set up INPO and their more effective use of EPRI. Second, utilities operating units to the highest standards will be damaged by the actions of incompetent utilities. Public confidence in *all* plant can be harmed by an incident at one plant and regulators will feel obliged to frame restrictive regulations designed to protect against the worst utilities.

A third failing, that utilities grossly overestimated their plant needs, is undeniable although hardly unique to US utilities. The degree of over-estimation would appear to have been not much worse than elsewhere although, in many cases, the realisation and subsequent adjustment process have been painfully slow.

Many of the weaknesses that can be attributed to the vendors arose from the rapid commercialisation process. It is arguable that vendors were not sufficiently responsible as salesmen. They had a responsibility to alert utilities to the scale of the task they were taking on (if indeed the vendors themselves appreciated it) and also to consider whether large nuclear power stations were the most appropriate solution to the utilities' needs. General Electric and Westinghouse also made serious errors in materials choice and water chemistry which have led to the problems of stress corrosion cracking and steam generator corrosion. It is perhaps significant that Babcock & Wilcox and Combustion Engineering have been more successful in this respect (although their steam generator designs have not been without flaws). From their long experience of boiler manufacture, they would be expected to be more skilled in specifying appropriate materials and conditions for pipe-work carrying boiling water.

One of the most obvious differences between the USA and the other countries examined is the use of third parties (outside the utility and vendor) to carry out the architect-engineering and construction functions. Many US utilities have strong links with favoured partners for major projects and they naturally turned to these partners to carry out nuclear projects. Architect-engineers/constructors should bring a number of relevant skills to such a project which a small utility might not be expected to possess. These include:
– experience with high technology, 'state-of-the-art' projects
– ability to manage large construction sites efficiently
– experience with component procurement

These advantages have, however, often been offset. Orders have been spread over a number of architect-engineers/constructors and learning may not have been effective especially as architect-engineers and constructors are unlikely to have wanted to pool their experience. For similar reasons of protecting their market niche, they are also unlikely to have been enthusiastic promoters of standardisation (unless the standard design was their own).

In addition, given the lack of further orders, architect-engineers and constructors have not had the same incentive to complete orders that a utility or vendor would have. For a utility, delays in construction cause heavy costs and, since they will be the operators, there is an incentive to maintain high quality. For a vendor, which will be judged on its product, there is an equally strong incentive to carry out the work quickly and efficiently.

Whilst an architect-engineer or constructor would, all things being equal, want to be associated with a successful project, given the complexity of the problems facing nuclear power it is unlikely that being associated with a failed or problematic project would hurt them as much as the utility or vendor. In fact, whilst design and regulatory changes have cost utilities dearly, architect-engineers have been able to charge on a 'cost-plus' basis since they were not anticipated when the contracts were placed. In that narrow sense, architect-engineers have done very good business from design changes.

Nevertheless it would be facile to see these arguments as a blanket condemnation of the architect-engineer/constructor system. Utilities in the USA which carry out their own architect-engineering have not universally had good experience and there are examples which prove that the use of a good architect-engineer, properly managed by the utility, can prove successful.

Regulators, both economic and safety, are also portrayed as the villains of the piece particularly from within the reactor supply industry. However, whilst there have been errors on this side, they do not appear to have been at the root of the problems. As has been argued previously, once the initial error of allowing the flood of orders in the mid-1960s through with inadequate scrutiny had been made, the AEC and its successor the NRC were in an impossible situation. This error was exacerbated by the failure of the utilities to ensure that high standards of work practice were applied *throughout* the industry. The NRC has responded to these variable standards by imposing fines which, although apparently large (sometimes of the order of $0.5m), are dwarfed by the replacement power costs for a shutdown lasting only a week. Fines may thus be seen as an irritant, damaging the public relations of the utility rather than a strong incentive to improve working practice. The other more significant reaction has been to increase the stringency of regulations, crudely, to make them 'idiot-proof', and in so doing to require a great deal of backfitting. The NRC can probably be justifiably criticised for not being sufficiently active in calling attention to the scale of the problem at an early stage and in promoting good practice.

There have also been serious criticisms about the lack of direction and unity in the NRC, perhaps stemming in part from the system of commissioners which oversees this body. The commissioners are frequently characterised as being too political and arguing from entrenched positions rather than responding flexibly to the current situation. It is undeniable that the commissioners do disagree with each other publicly and whilst this disagreement may offer a worthwhile plurality, the disunity may percolate throughout the organisation. In practical terms, the result has been that the treatment of backfits has been piecemeal, with too little co-ordination to consider the systems impact of a variety of backfits and whether they represent the most cost-effective method of improving the safety of plants.

Criticisms of the economic regulators centre on the treatment of the costs of plant under construction and, more generally, the extent to which utilities should be able to pass on whatever costs they incur to consumers. In order to put these criticisms in perspective, it is worth examining the historical context. Up until the late 1960s, utilities had frequently been able to reduce the cost of electricity not just in real terms but in nominal dollar terms. This consistent reduction in prices was made possible by the flow of technical progress in generation and transmission technology and by a consistent fall in the cost of fossil-fuels. Electricity demand was growing steadily at 7% per annum or more requiring utilities to order new plant to keep up with demand. This meant that there was little pressure on PUCs to carry out stringent reviews of utility operation and the requirement that utility investment in new supply capacity be shown to be 'used and useful' was not a significant factor.

Electric utilities were thus seen as exceedingly safe investments since their markets were protected and they were effectively guaranteed a return on the full cost of any investment they made. Indeed it can be argued that because regulation of electricity prices is on a rate-of-return basis, i.e. the utility is allowed to make a specified return on its asset base, there was an incentive for utilities to overinvest in plant and other equipment to give them a larger capital base on which to make a return.

The situation began to change as the rate of technical progress slowed, the cost of fossil-fuels began to rise and utilities started to raise electricity rates. This resulted in pressures on PUCs to adopt a much tougher approach to electric utilities. With many of the PUC commissioners ambitious political appointees, there were strong incentives to curb these increases. From the mid- to late 1970s onwards, the combination of increasing fossil-fuel costs, lower than expected electricity demand growth and rapidly escalating capital costs of all plant, particularly nuclear plant, meant that the hurdle of plant being 'used and useful' was a real one. Utilities began to lose their 'blue chip' status raising their cost of borrowing especially if they had a programme of plant construction.

Whether the pendulum has swung too far against utility investments, particularly for capital intensive projects, is arguable, but it seems clear that the previous situation, where utility investments were almost risk-free to the utility, was not necessarily conducive to the tight control of costs. However, if utilities can demonstrate good control of their projects and sound planning and decision-making, there would seem to be no intrinsic reason why the present system should not respond appropriately.

INSTITUTIONAL REFORM AND PROSPECTS FOR NEW ORDERS

The American nuclear industry faces difficult questions about the direction of future market and technology trends. These difficulties can be summarised under three headings:
– lack of home orders
– divergence from the world mainstream
– loss of technological capability

Lack of home orders. The prospect of new orders for nuclear power plant seems further away than ever. This arises partly because the considerable over-ordering of plant of all types in the late 1960s and early 1970s has, even allowing for cancellations, left the USA with a surplus of plant that may persist into the 1990s, and partly because of the financial community's aversion to plant construction programmes especially for nuclear plant. For a utility faced with an apparent shortage of capacity, building a nuclear plant would be rated as a very poor fourth choice behind a conservation programme, wheeling in power from adjacent utilities and building a coal-fired plant. Even the market to replace old power stations, previously expected to become important in the 1990s, now looks in doubt from the current trend towards cost-effective refurbishment.

Divergence from the world mainstream. Many utilities (with the notable exceptions of Duke Power and Commonwealth Edison) believe that future orders should be for much smaller plant (400–600 MW) than in the past. This is partly because smaller plant is seen as having a smaller economic risk due to its lower capital cost and potentially shorter lead-time; for a small utility with low demand growth a large plant may represent too 'lumpy' an addition to its asset base; for some utilities there is a strong belief that the scaling-up process went too far and too fast and that smaller plant is intrinsically likely to perform better. Were the USA to follow this path, it would be diverging markedly from the world trend where there is little evidence yet of any dissatisfaction with 900–1200 MW units.

Loss of technological capability. As has been noted earlier the loss of a significant proportion of manufacturing capability is inevitable and has already started. However, given the current and probable future world over-capacity in supply of all major components, this is probably of less significance than the loss of design capability. Increasingly, technology development is taking place outside the USA, in West Germany, France and Japan. Westinghouse and General Electric are attempting to keep in touch by entering into joint ventures with their licensees in Japan but, in the absence of home orders, even this strategy cannot prove effective indefinitely.

Possible courses

Two of the options that have been suggested for breaking out of the current impasse are, firstly, bringing forward and perhaps subsidising new orders ahead of need and, secondly, developing new reactor types such as PIUS (a Swedish concept) or the high-temperature reactor.

The first of these seems undesirable even in the unlikely event that the necessary finance could be raised since one or two orders will do little to maintain capabilities even for one or two vendors. Until the fundamental problems of the industry are resolved, it seems unwise to embark on a path which may repeat the errors of the past. The development of other reactor types has some attractions but many of the industry's problems are institutional and would not be solved by simply changing the technology. In addition, the option seems unlikely to be feasible since no utility could currently take the risk of ordering even the reasonably proven LWR, let alone an unproven reactor type, which the high-temperature reactor and PIUS essentially are.

A more prudent course would seem to be that signposted by EPRI and INPO. The loss of a certain amount of capability should be regarded as inevitable and priority must be placed on the utility industry putting its house in order. In practice this means demonstrating the maturation of the technology by completing the backlog of orders as near to time and cost as possible and ensuring that plant is operated to high standards, achieving good capacity factors and avoiding incidents which cast doubt on the safety of the technology.

For future orders, much of this current debate may prove academic given the length of time before orders are feasible. As was amply demonstrated by Chernobyl, external events and changed perceptions, impossible to predict at this stage, might rapidly overtake such deliberations. Nevertheless, if the debate focuses on the role of the key institutions, bringing out their past failings, it will leave the industry in a healthier state.

EPRI has already made an important contribution to improving reactor performance not only with its work on identifying the causes and remedies for stress corrosion cracking in BWRs and steam generator corrosion in PWRs but in a variety of smaller programmes. For the future, there seems to be a change in direction as these 'fire-fighting' programmes are completed towards programmes aimed at positively influencing the direction of technology development. In particular EPRI is in the process of launching programmes looking at standardisation and unit size.

Standardisation is a widely abused term, covering a range of concepts from plant that is identical in all respects including component suppliers, and lay out, for example the French programme, to plant whose sub-systems are regarded as 'black boxes' which are defined in terms of their performance not the equipment that makes them up. There are dangers of technological stagnation with the former approach and, for the USA, it is not a practical option given anti-trust laws. The latter approach, which is being followed by EPRI, has the merit of flexibility and is user-specified. However, the EPRI programme is only part of the answer and it is necessary that the vendor, of whatever nationality, carry out the com-plementary R&D necessary to produce the equipment going into these 'black boxes'.

EPRI is also launching a programme to examine the design of small units for future orders. A prior step may be to understand more fully whether there is a size effect and how it arises. There is a danger that if, for example, there is a size effect which derives from complexity, orders for small plant will be met simply using scaled-down large units. In this situation utilities will be left with something which is no less complex than a large unit but which loses any scale economies.

On the manufacturing side, as has been argued previously, the main difficulty facing vendors, if they are to stay in the market, is what resources to maintain and how to maintain them. In addition, vendors must decide whether to take firmer control of projects by taking on more of the architect-engineering function. The project that Westinghouse launched in the mid-1970s to produce factory-built units mounted on barges which could be floated into position was a bold but still-born move in this direction, whilst the current standardised designs such as General Electric's GESSAR, Combustion Engineering's CESSAR and Westinghouse's SNUPPS are less radical steps. Overall, however, the most critically important moves are those being instigated by the utilities to take control of the direction of development of the technology.

APPENDIX. DETAILED ANALYSIS OF PLANT PERFORMANCE OF US PLANT

The analyses in this appendix amplify the analyses of plant performance given in the main text giving greater detail on the causes of lost output and reliability of the plant systems and include all operating data accumulated up to the end of 1984. Four tables are presented for each vendor. In the first and second tables, causes of lost output – shutdown, derating and operating losses – are shown as a percentage of total potential output. The 'adjusted operating losses' normalise operating losses to those which would have occurred over a year of continuous operation. This adjustment takes account of the fact that operating losses can only occur when the unit is producing power. Thus a unit which is shutdown for much of the year will, by definition, have low adjusted operating losses which will tend to understate the significance of below-capacity operation. The first table categorises causes of lost output by age of reactor and illustrates the maturation process identifying what problems are characteristic of new units and what problems are persistent. The second table is split up into three time periods – pre-TMI (up to 1979), 1979, and post-TMI (1980–84). This table includes only the performance of mature units (those with more than two years commercial operation) and shows the impact of the TMI accident and whether any overall technology maturation is evident in operating performance.

The third and fourth tables subdivide shutdown losses according to reason for the shutdown. The data shown give the notional number of hours a unit would have been shut down for the specified cause to complete 8760 hours (1 year) of actual operation. The shutdown frequency gives the notional number of shutdowns a unit would have undergone to complete 8760 hours of actual operation. Notional figures are presented rather than actual number of hours lost and actual number of shutdowns, to reflect the fact that a unit can only be shutdown when it is operating. Thus, a unit which is shutdown for most of the year will inevitably have low unadjusted failure rates and shutdown frequencies, which do not reflect the true reliability of the systems.

The third table corresponds to the first table being split according to unit age whilst the fourth table corresponds to the second table being split according to time period.

Table A5.1. *Causes of lost output in Westinghouse units according to age (%)*[a]

Age	No. of reactor years	Capacity factor	Shutdown losses	Derating losses	Operating losses	Adjusted operating losses
2-loop units						
1	4	59.9	21.8	0.8	17.4	23.2
2	4	70.4	18.9	0.9	9.7	12.2
3 onwards	59	76.0	17.1	1.5	5.5	6.8
3-loop units						
1	7	58.0	33.4	0.8	7.8	12.9
2	9	57.3	33.9	1.6	7.1	11.6
3 onwards	67	57.8	34.4	3.2	4.6	8.1
4-loop units						
1	12	58.9	26.6	2.8	11.7	16.0
2	12	45.7	44.7	2.3	7.3	15.3
3 onwards	56	56.6	34.7	1.3	7.3	12.7

[a] Losses are expressed as a percentage of potential output.
Source: SPRU nuclear power plant data base.

Table A5.2. *Causes of lost output in mature Westinghouse units according to time period*[a]

Time period	No. of reactor years	Capacity factor	Shutdown losses	Derating losses	Operating losses	Adjusted operating losses
2-loop units						
1973–78	23	76.7	16.4	−0.3	7.2	8.9
1979	6	75.6	18.4	2.4	3.6	4.5
1980–84	30	75.5	17.3	2.7	4.6	5.5
3-loop units						
1974–78	19	67.4	25.2	2.1	5.3	7.0
1979	6	40.2	51.5	3.4	4.9	10.6
1980–84	42	55.9	36.1	3.7	4.2	8.3
4-loop units						
1976–78	8	64.7	26.2	1.0	8.1	11.1
1979	6	57.8	33.5	1.1	7.6	11.3
1980–84	42	54.9	36.5	1.5	7.1	13.2

[a] Losses are expressed as a percentage of potential output.
Source: SPRU nuclear power plant data base.

Table A5.3. *Causes of shutdown in Westinghouse units according to age*

Age	No. of reactor years	Shutdowns (hours per year of operation)																Shutdown frequency (no. per year of operation)
		1	3	4	5	6	7	8	9	10	B	C	D	E	F	G	H	
2-loop units																		
1	4	47	9	0	196	6	139	1256	112	6	0	0	617	79	0	0	0	19.1
2	4	83	13	3	117	17	7	161	38	0	0	1149	163	60	7	0	7	7.7
3 onwards	59	12	7	16	16	125	32	102	21	0	3	1084	181	4	3	39	8	5.6
3-loop units																		
1	7	0	31	67	543	25	331	807	223	6	52	533	1211	11	19	5	42	20.2
2	9	0	29	13	284	284	27	97	513	19	79	1945	393	13	0	0	50	14.8
3 onwards	67	1	38	25	263	308	68	116	109	16	30	1946	630	10	1	216	37	14.2
4-loop units																		
1	12	1	74	20	295	78	195	644	128	27	38	637	616	125	0	2	39	16.1
2	12	14	131	151	223	236	104	884	8	35	60	2326	738	108	12	795	218	24.3
3 onwards	56	1	74	89	204	124	129	599	64	27	47	1976	298	20	0	153	123	17.0

Causes of shutdown are:

1 = reactor and accessories
2 = fuel
3 = reactor control system and instrumentation
4 = nuclear auxiliary and emergency system
5 = main heat removal system
6 = steam generators
7 = feedwater, condenser and circulating water systems
8 = turbine generator
9 = electrical power and supply system
10 = miscellaneous equipment failure

B = operating error
C = annual refuelling and maintenance shutdown
D = maintenance and repair
E = testing of plant systems/components
F = training and licensing
G = regulatory limitation
H = other

Source: SPRU nuclear power plant data base.

Table A5.4. *Causes of shutdown in mature Westinghouse units according to time period*

Time period	No. of reactor years	Shutdowns (hours per year of operation)																Shutdown frequency (no. per year of operation)
		1	3	4	5	6	7	8	9	10	B	C	D	E	F	G	H	
2-loop units																		
1973–78	23	31	15	35	19	160	13	215	42	0	1	951	78	8	4	1	13	6.7
1979	6	0	0	0	34	276	57	149	0	0	0	879	370	0	0	152	14	6.6
1980–84	30	0	3	4	11	69	42	6	10	0	4	1226	222	0	4	45	3	4.6
3-loop units																		
1974–78	19	2	15	47	146	678	18	48	9	17	0	1385	222	10	4	0	0	9.9
1979	6	0	27	40	376	75	36	46	0	0	0	2904	80	62	0	1693	58	13.3
1980–84	42	0	49	13	300	173	95	156	170	18	48	2062	893	3	0	102	50	16.3
4-loop units																		
1976–78	8	0	6	3	255	54	139	39	11	5	16	1742	299	33	0	0	14	10.6
1979	6	0	16	11	18	54	53	21	23	0	96	1700	547	6	0	915	84	14.0
1980–84	42	2	96	117	220	148	138	788	79	35	45	2060	262	19	0	74	150	18.7

Note: See Table A5.3 for a list of the causes of shutdown.
Source: SPRU nuclear power plant data base.

Table A5.5. *Causes of lost output in General Electric units according to age*[a]

Age	No. of reactor years	Capacity factor	Shutdown losses	Derating losses	Operating losses	Adjusted operating losses
500–700 MW						
1	3	54.6	23.6	1.8	19.9	27.0
2	5	50.4	38.4	−0.6	11.8	19.3
3 onwards	73	61.2	28.8	1.8	8.2	14.2
700–950 MW						
1	8	56.3	24.9	0.7	18.0	23.9
2	9	50.6	30.4	2.6	16.4	24.4
3 onwards	78	56.2	29.4	2.2	12.2	17.5
950+ MW						
1	6	44.7	41.0	1.0	13.3	23.5
2	5	52.8	36.5	0.8	10.0	19.4
3 onwards	36	59.6	32.7	0.9	6.9	11.4

[a] Losses are expressed as a percentage of potential output.
Source: SPRU nuclear power plant data base.

Table A5.6. *Causes of lost output in mature General Electric units according to time period*[a]

Time Period	No. of reactor years	Capacity factor	Shutdown losses	Derating losses	Operating losses	Adjusted operating losses
500–700 MW						
1973–78	31	65.2	23.5	1.6	9.7	13.3
1979	7	74.7	17.1	1.8	6.3	8.0
1980–84	35	54.9	35.8	2.0	7.2	16.2
700–950 MW						
1973–78	22	59.7	22.6	1.0	16.7	22.1
1979	8	57.8	26.3	3.1	12.8	17.2
1980–84	48	54.3	33.1	2.6	10.0	15.5
950+ MW						
1977–78	7	59.7	27.5	1.2	11.7	16.5
1979	4	79.4	12.6	1.0	7.0	8.1
1980–84	25	56.4	37.3	0.8	5.5	10.5

[a] Losses are expressed as a percentage of potential output.
Source: SPRU nuclear power plant data base.

Table A5.7. *Causes of shutdown in General Electric units according to age*

Age	No. of reactor years	Shutdowns (hours per year of operation)																Shutdown frequency (no. per year of operation)
		1	2	3	4	5	7	8	9	10	B	C	D	E	F	G	H	
500–700 MW																		
1	3	0	873	17	150	245	29	157	122	8	11	520	73	290	5	27	5	21.5
2	5	664	0	162	0	226	4	119	10	42	3	641	1011	69	0	1104	31	19.1
3 onwards	73	41	0	21	157	265	75	70	62	107	50	1674	281	17	3	39	68	13.9
700–950 MW																		
1	8	1	0	38	177	504	82	200	9	52	20	623	700	137	0	77	99	22.2
2	9	7	0	17	124	359	180	89	37	24	106	1966	255	126	0	95	14	18.6
3 onwards	78	0	0	32	92	144	93	185	73	62	60	1862	395	4	0	88	52	14.8
950+ MW																		
1	6	9	0	53	24	164	203	230	30	33	2234	335	472	0	0	167	281	20.0
2	5	0	0	270	94	274	96	332	52	11	1344	1044	100	7	0	0	31	17.5
3 onwards	36	3	0	21	145	145	68	116	58	244	143	1993	336	28	0	5	105	17.6

Note: See Table A5.3 for a list of the causes of shutdown.
Source: SPRU nuclear power plant data base.

Table A5.8. Causes of shutdown in mature General Electric units according to time period

Time period	No. of reactor years	Shutdowns (hours per year of operation)															Shutdown frequency (no. per year of operation)
		1	3	4	5	7	8	9	10	B	C	D	E	F	G	H	
500–700 MW																	
1973–78	31	21	19	51	85	143	87	109	209	46	1305	203	8	7	46	56	10.5
1979	7	0	9	48	371	17	31	0	6	26	827	75	4	0	0	227	7.6
1980–84	35	68	25	274	402	26	63	34	36	59	2169	393	27	0	40	47	18.1
700–950 MW																	
1973–78	22	0	32	148	137	118	140	49	13	40	1342	166	3	0	80	53	13.5
1979	8	0	5	26	51	21	119	329	0	69	1179	256	6	0	558	41	10.7
1980–84	48	0	36	78	164	93	217	41	96	68	2214	523	4	0	14	53	16.1
950+ MW																	
1977–78	7	18	22	93	76	81	97	12	31	0	2116	106	11	0	0	57	14.9
1979	4	0	9	0	121	22	48	6	0	0	688	313	6	0	0	39	8.9
1980–84	25	0	23	183	169	71	133	79	342	206	2168	404	36	0	7	129	19.7

Note: See Table A5.3 for a list of the causes of shutdown.
Source: SPRU nuclear power plant data base.

Table A5.9. *Causes of lost output in Babcock & Wilcox units according to age*[a]

Age	No. of reactor years	Capacity factor	Shutdown losses	Derating losses	Operating losses	Adjusted operating losses
1	7	53.4	37.8	1.9	6.8	11.5
2	8	55.8	36.1	2.6	5.4	9.7
3 onwards	51	59.9	30.8	3.1	6.2	9.5

[a] Losses are expressed as a percentage of potential output.
Source: SPRU nuclear power plant data base.

Table A5.10. *Causes of lost output in mature Babcock & Wilcox units according to time period*[a]

Time Period	No. of reactor years	Capacity factor	Shutdown losses	Derating losses	Operating losses	Adjusted operating losses
1976–78	12	65.0	25.4	3.0	6.7	9.3
1979	5	59.6	32.4	3.3	4.7	6.9
1980–84	34	58.1	32.6	3.1	6.2	10.0

[a] Losses are expressed as a percentage of potential output.
Source: SPRU nuclear power plant data base.

Table A5.11. *Causes of shutdown in Babcock & Wilcox units according to age*

Age	No. of reactor years	Shutdowns (hours per year of operation)																Shutdown frequency (no. per year of operation)
		1	2	3	4	5	6	7	8	9	10	B	C	D	E	G	H	
1	7	3	709	349	76	219	73	64	922	175	24	12	0	1088	58	366	118	18.5
2	8	164	0	222	35	414	109	260	58	0	356	14	2012	426	0	0	2	14.4
3 on wards	51	78	0	112	58	184	419	56	286	32	16	11	1687	238	21	239	72	11.7

Note: See Table A5.3 for a list of the causes of shutdown.
Source: SPRU nuclear power plant data base.

Table A5.12. *Causes of shutdown in mature Babcock & Wilcox units according to time period*

Time period	No. of reactor years	Shutdowns (hours per year of operation)															Shutdown frequency (no. per year of operation)
		1	3	4	5	6	7	8	9	10	B	C	D	E	G	H	
1976–78	12	199	122	42	150	516	23	142	2	44	18	1375	172	27	0	18	12.2
1979	5	93	82	0	141	168	31	506	10	0	4	1031	79	133	1449	354	14.2
1980–84	34	33	113	72	202	422	72	304	46	8	10	1893	284	3	145	49	11.2

Note: See Table A5.3 for a list of the causes of shutdown.
Source: SPRU nuclear power plant data base.

Table A5.13. *Causes of lost output in Combustion Engineering units according to age*[a]

Age	No. of reactor years	Capacity factor	Shutdown losses	Derating losses	Operating losses	Adjusted operating losses
1	9	63.5	21.9	5.2	9.5	12.4
2	7	59.7	28.2	5.2	7.0	10.2
3 onwards	45	68.2	24.9	2.7	4.3	5.7

[a] Losses are expressed as a percentage of potential output.
Source: SPRU nuclear power plant data base.

Table A5.14. *Causes of lost output in mature Combustion Engineering units according to time period*[a]

Time Period	No. of reactor years	Capacity factor	Shutdown losses	Derating losses	Operating losses	Adjusted operating losses
1975–78	8	68.7	20.3	6.2	4.9	6.1
1979	5	66.8	24.9	2.8	5.5	7.6
1980–84	32	68.3	26.0	1.8	3.9	5.4

[a] Losses are expressed as a percentage of potential output.
Source: SPRU nuclear power plant data base.

Table A5.15. *Causes of shutdown in Combustion Engineering units according to age*

Age	No. of reactor years	Shutdowns (hours per year of operation)																Shutdown frequency (no. per year of operation)
		1	3	4	5	6	7	8	9	10	B	C	D	E	F	G	H	
1	9	10	42	18	180	146	153	174	108	88	28	1027	176	100	4	37	139	19.8
2	7	96	157	51	120	8	311	9	79	65	10	1487	675	30	0	0	47	11.1
3 onwards	45	16	39	17	186	35	81	60	40	13	42	1704	107	5	1	107	43	10.0

Note: See Table A5.3 for a list of the causes of shutdown.
Source: SPRU nuclear power plant data base.

Table A5.16. *Causes of shutdown in mature Combustion Engineering units according to time period*

Age	No. of reactor years	Shutdowns (hours per year of operation)																Shutdown frequency (no. per year of operation)
		1	3	4	5	6	7	8	9	10	B	C	D	E	F	G	H	
1975–78	8	0	20	0	113	3	86	87	67	12	9	1546	80	20	5	0	21	11.1
1979	5	37	12	0	254	231	5	113	3	3	0	1132	202	7	0	528	159	7.4
1980–84	32	17	48	24	193	12	92	45	39	14	57	1833	100	0	0	68	30	10.2

Note: See Table A5.3 for a list of the causes of shutdown.
Source: SPRU nuclear power plant data base.

CHAPTER 6

Federal Republic of Germany

Introduction

Although the Federal Republic of Germany's entry into the field of nuclear power technology was delayed by post-war restrictions on technology with military applications, a number of factors enabled it to catch up rapidly with its competitors and make a significant impact on technology development. In particular, there was:

- a number of companies, especially Siemens, which were world leaders in heavy electrical engineering
- a large pool of high quality engineering and metallurgical skills which are vital to successful nuclear power development
- a number of scientists who had played a key role in elucidating basic nuclear physics
- no overt connection with any military applications which in other countries may have distorted civil developments.

To set against these advantages, public opposition to nuclear power has been particularly strong delaying the construction of a number of plants, severely curtailing development of the fast breeder reactor and virtually halting development of the back-end of the fuel cycle. Despite this, however, FR Germany was the first country importing light water technology to achieve independence from the USA and it is arguable that German technology is now more mature than that of its competitors. In addition, its supply industry looks equipped to survive the difficult market conditions that seem likely to apply for the next decade.

In this chapter we examine what factors lay behind the technological success and what the impact of the strong, sophisticated opposition movement has been.

THE ENERGY AND ECONOMIC BACKGROUND

FR Germany has been one of the most successful post-war economies in the world. Over the period 1960–84, GDP grew at 3.3% per annum although, as with most other economies, the two oil shocks have dampened down growth – from 1960 to 1973 growth was 4.4% per annum, but from 1973 to 1979, growth had fallen to 2.4% per annum and to less than 1% from 1979 to 1984. Again, as elsewhere, manufacturing production suffered more severely than GDP as a whole during the two recessions with output falling by about 6% in each case.

On the energy side (see Table 6.1 and Table 6.2), although FR Germany has consistently met about 50% of its primary energy needs through domestic production, about 70% of this production is coal or lignite with relatively little hydro-carbon or hydro-electric production. Over the past decade coal and lignite production have declined by about 10%, but the growth in nuclear power has more than compensated for this fall. Increasingly, solid fuels have been used for electricity generation and the growth in demand in this sector over the past 10 years has been met almost equally by increased use of solid fuels and nuclear power. In 1984, nuclear power accounted for about 22% of electricity production and the completion of a number of new units in 1983/84 resulted in nuclear production increasing its share in 1985. With a number of further units close to completion, the nuclear share of electricity generation may have risen to about 40% by 1990. Gas purchased under long-term contracts from the Netherlands continues to be used in power stations, although non-power station uses are increasingly being supplied by imports from the Soviet Union.

On the demand side, only in the transport sector is demand still consistently rising and industry's fuel demands have fallen 20% (mainly coal and oil) since 1973. Nevertheless, electricity demand has increased in all sectors with an overall average growth rate of about 2% per annum. Growth has been particularly strong in the 'residential' and 'other' (primarily public services and commerce) sectors.

THE DEVELOPMENT OF NUCLEAR POWER

The development of nuclear power in FR Germany can be conveniently split into five phases.

The first phase covers the period up to 1962 when the first order for a commercial-size reactor, the 237 MW Gundremmingen BWR, was placed. The effect of this was somewhat similar to that of the Oyster Creek order in the USA and, during the second phase (1962–69), the focus of technology development narrowed concentrating on LWRs and the firms supplying them. AEG supplied BWRs under licence from GE (USA) and Siemens supplied PWRs under licence from Westinghouse (USA). The impact was not, however, as abrupt as in the USA and work continued on a number of other reactor designs such as heavy water cooled, gas-cooled, fast breeder and high-temperature reactors. By 1969 AEG's financial problems were

Table 6.1. *Production and consumption of primary energy in FR Germany – 1973–84 (million tonnes of oil equivalent)*

	Production			Consumption			Consumption for electricity generation		
	1973	1978	1984	1973	1978	1984	1973	1978	1984
Solid fuels	96.3	87.1	84.1	87.1	76.8	85.3	48.0	52.7	59.4
Liquid fuels	6.7	5.2	5.5	148.4	143.3	111.6	6.3	6.1	1.8
Natural gas	15.4	16.1	14.3	27.6	41.5	41.4	7.5	14.2	7.2
Nuclear power	2.6	8.0	20.7	2.6	8.0	20.7	2.6	8.0	20.7
Other primary electricity	3.5	4.1	4.1	3.5	4.1	4.1	3.5	4.1	4.1
Total	124.5	120.5	128.7	269.2	273.7	263.1	67.9	85.1	93.2

Source: International Energy Agency, *Energy Balances of OECD Countries*, various, Paris.

Table 6.2. *Delivered energy consumption in FR Germany – 1973–84 (million tonnes of oil equivalent)*

	Industry			Transportation			Residential[a]			Other[a]			Total		
	1973	1978	1984	1973	1978	1984	1973	1978	1984	1973	1978	1984	1973	1978	1984
Solid fuels	24.1	16.3	18.4	0.7	0.1	—	9.1	4.4	3.7	1.0	0.9	1.6	34.9	21.7	23.7
Liquid fuels	36.6	28.6	20.9	32.3	38.2	41.5	48.6	30.6	22.8	3.1	20.2	14.0	120.6	117.6	99.2
Gas	14.0	13.0	16.0	—	—	—	5.0	8.5	12.1	2.8	4.7	5.7	21.8	26.2	33.8
Electricity	12.2	13.2	13.9	0.8	0.8	0.9	5.2	6.9	8.1	4.0	5.3	6.4	22.2	26.2	29.3
Total	86.9	71.1	69.2	33.8	39.1	42.4	67.9	50.4	46.7	10.9	31.1	27.7	199.5	191.7	186.0

[a] The allocation of energy demands between the 'residential' and 'other' sectors for 1973 does not appear to be satisfactory.

Source: International Energy Agency, *Energy Balances of OECD Countries*, various, Paris.

such that it was forced to amalgamate its reactor and large turbine generator interests with Siemens to form Kraftwerk Union (KWU), marking the end of the second phase.

The third phase (1969–75) saw a fairly steady flow of orders for full-size (larger than 1000 MW) units, as the German industry became fully independent of its US licensors. Development of alternative reactor designs narrowed to the fast breeder and high-temperature reactors and public opposition to nuclear power became increasingly important. Export markets began to be built up particularly in neighbouring European countries and in South America.

The fourth phase (1975–79) started with plans for a massive expansion of nuclear power on the scale of the French programme which, for a number of reasons, notably low electricity demand growth, were not realised. Opposition became increasingly sophisticated, switching from physical opposition at sites to opposition through the courts. By 1979, the process of building nuclear power plant seemed to be hopelessly entangled and the nuclear option at risk of being lost. In technology terms, the two reactor types were faring differently. The PWRs were performing extremely well and in important respects seemed to be technologically superior to the US parent design. By contrast, the major pipe-work of the BWRs was judged inadequate to meet the requirements of the newly derived 'Basis Safety Concept' (see later). The two earliest units, where severe pipe-work cracking had been found, were prematurely retired and extended pipe-work replacement work lasting about a year was carried out at the other units.

From 1979 onwards (the fifth phase), the vendor, KWU, and the utilities have been struggling, with some success, to find ways of streamlining the licensing process with the so-called 'convoy system' (see later for a full description). Public opposition to nuclear power seems to have become less vigorous and the German supply industry, which has never suffered the bulges in ordering that have distorted the US and French industries, now appears better equipped to survive the next decade than some of its competitors. Technologically, the PWR's achievements remain impressive whilst there are signs, both in the performance of repaired BWRs and in newer BWRs which have avoided the cracking problem, that the BWR may re-emerge as a strong competitor.

THE KEY INSTITUTIONS

The institutional structure is complex and is somewhat similar to that found in the USA. As in the USA, the institutions can be divided into four groups, the electricity supply industry, the plant supply industry, Government (both federal and regional) and regulators. In addition, there are important differences between the political parties which have had a significant effect in some regions.

The electricity supply industry

At first sight the utility structure appears extremely complex with several hundred electric utilities. However, closer examination shows that there are a small number of utilities that own and control a high proportion of total capacity. German electricity producers can be divided into four categories:
- the eight regional, nationally co-ordinated utilities
- 41 smaller regional utilities
- local (often municipal) utilities
- industrial producers

The eight nationally co-ordinated utilities* have legally-protected catchment areas within which they operate the grid. Some of these utilities sell electricity directly to final consumers whilst others, such as Bayernwerk, sell most of their output to distributing companies. There is no national grid as such, but seven of the eight utilities are interconnected (the eighth, Berliner Kraft und Licht, is geographically isolated) and transfers of power do take place. These seven interlinked utilities (listed in Table 6.3) own most of the nuclear plant in operation and under construction and also meet regularly to plan grid development.

A number of these utilities have substantial interests beyond electricity supply. For example Preussenelektra, the second largest electric utility, is part of VEBA which is also Germany's largest oil company and fourth largest chemical company (Financial Times, 1985). RWE has substantial holdings in nuclear fuel cycle companies (e.g. Nukem), service companies (Lahmeyer International), other electric utilities (Isarwerke) and a coal company (Rheinische Braunkohlenwerke).

Nearly all the rest of the nuclear capacity is owned by the regional electricity producers and the local producers (of which there are about 1000). The industrial producers (including the German railway) own about one-sixth of total installed capacity of all types and generally produce electricity primarily for their own use. However, some of the mining companies such as STEAG and Saarbergwerke do produce substantial quantities of electricity for sale to the grid. Although most electric utilities are private companies, the vast majority have at least 25% of their stock held publicly by local governments. For example local governments own 30% of RWE and control 60% of voting rights (Nucleonics Week, 1985b, p. 9).

---

* In 1985 Preussenelektra merged with its former subsidiary Nordwestdeutsche Kraftwerke (NWK), reducing the number of such utilities from nine to eight.

Table 6.3. *Ownership of nuclear power plant in FR Germany*

	Nuclear capacity[b] (MW)	
Utilities[a]	In operation	Under construction
RWE	4 464	1 308
NWK/Preussenelektra	3 903	1 092
Bayernwerk	2 407	548
HEW	1 394	273
EVS	1 253	260
Badenwerk	1 229	0
VEW	0	975
Others	2 085	2 512
Total	16 785	6 967

[a] *Utilities:*
   RWE – Rheinisch-Westfalisches Elektrizitatswerk AG
   Preussenelektra – Preussenelektra AG
   NWK – Nordwestdeutsche Kraftwerke AG
   Bayernwerk – Bayernwerk AG
   HEW – Hamburgische Electricitatswerke AG
   EVS – Energie Versorgung Schwaben AG
   Badenwerk – Badenwerk AG
   VEW – Vereinige Elektrizitatswerke Westfalen AG
[b] Capacities are as of 1 January 1985.
Source: SPRU nuclear power plant data base.

Most nuclear plant is operated by companies which are separate legal entities, for example Philippsburg is nominally owned by Kernkraftwerk Philippsburg but in practice these companies are fully under the control of the utility with the largest shareholding in the project.

Two other institutions that should be mentioned are the Vereinigung Deutscher Elektrizitatswerke (VDEW) and the Technische Vereinigung der Grosskraftwerksbetreiber (VGB). The VDEW acts partly to represent the utilities' interests to the public but also as a clearing house for commercial information of relevance to the utilities. The VGB, which includes representatives of all owners of large combustion plant is an association which commissions and carries out research on the operation of large combustion plants.

The plant supply industry

The plant supply industry has been based largely on the major traditional heavy electrical companies Siemens, AEG and Brown Boveri (BBC) (see Table 6.4 and Table 6.5), and has been underpinned by the large engineering groups such as Gutehoffnungshutte (GHH), Mannesmannrohren, Deutsche Babcock and Sulzer.

Table 6.4. *The German nuclear programme – prototypes*

Name	Type[a]	MW (gross)	Year of order	Start of construction	Year of commissioning	Year of decommissioning	Supplier
VAK	BWR	16	1958	1958	1961	1985	GE/AEG
MZFR	PHWR	57	1961	1961	1966	1984	Siemens
AVR	HTR	15	1961	1961	1967	1985	BBC/Krupp
Grosswelzheim	BWR	25	1963	1965	1970	1971	AEG
KNK I/KNK II[b]	FBR	21	1966	1966/73	1972/77	–	Interatom
Niederaichbach	GCHWR	100	1964	1966	1972	1974	Siemens
Uentrop	HTR	300	1971	1971	1986	–	HRB
Kalkar	FBR	295	1972	1973	–	–	Interatom

[a] Reactor types: BWR – Boiling water reactor
PHWR – Pressurised heavy water reactor
HTR – High temperature reactor
FBR – Fast breeder reactor
GCHWR – Gas-cooled heavy water reactor

[b] KNK I was designed to gain experience with sodium as a coolant, and used a moderator, zirconium hydride. After two years of operation it was closed and the core replaced with a 'fast' core, i.e. with no moderator.

Source: SPRU nuclear power plant data base.

Table 6.5. *The German nuclear programme – commercial plant[a]*

Name	Type	MW (gross)	Year of order	Start of construction	Commercial operation	Owners
Gundremmingen[b]	BWR	250	1962	1962	1967	Badenwerk, RWE
Lingen[b]	BWR	250	1964	1964	1968	VEW
Obrigheim	PWR	345	1964	1965	1969	EVS (35%), Badenwerk (28%), Others (37%)
Stade	PWR	622	1967	1967	1972	NWK (67%), HEW (33%)
Wurgassen	BWR	670	1967	1967	1975	Preussenelektra
Biblis A	PWR	1200	1969	1970	1975	RWE
Brunsbuttel	BWR	805	1969	1970	1977	HEW (67%), NWK (33%)
Philippsburg 1	BWR	900	1970	1970	1980	Badenwerk (50%), EVS (50%)
Unterweser	PWR	1300	1971	1971	1979	NWK (50%), Preussenelektra
Biblis B	PWR	1300	1971	1972	1977	RWE
Isar 1	BWR	907	1971	1972	1979	Badenwerk (50%), Isar-Amperwerke (50%)
Krummel	BWR	1300	1972	1974	1984	HEW (50%), NWK (50%)
Neckarwestheim 1	PWR	855	1972	1972	1976	Various
Mulheim Kaerlich	PWR	1308	1973	1975	1987	RWE
Gundremmingen B	BWR	1310	1974	1976	1984	RWE
Gundremmingen C	BWR	1310	1974	1976	1985	RWE
Grohnde	PWR	1300	1974	1976	1985	Preussenelektra (50%), Weser (50%)
Grafenrheinfeld	PWR	1300	1975	1975	1982	Badenwerk
Philippsburg 2	PWR	1365	1975	1977	1985	Badenwerk (50%), EVS (50%)
Brokdorf[c]	PWR	1365	1975	1981	1986	NWK (50%), Preussenelektra (30%), HEW (20%)[c]
Neckarwestheim 2	PWR	1300	1982	1983	1989	Various
Isar 2	PWR	1370	1982	1982	1988	Badenwerk (40%), Others (60%)
Emsland	PWR	1300	1982	1982	1989	VEW (75%), Elektromark (25%)

[a] A number of other units have been ordered or letters of intent have been issued, but no significant site work has yet taken place. These include Wyhl, Neupotz, Biblis C and Hamm. Of the units listed in the table the NSSS and turbine generator for all PWRs except Mulheim Kaerlich were supplied by Siemens/KWU. The NSSS for Mulheim Kaerlich was supplied by AEG/KWU whilst the NSSS and turbine generator for all BWRs were supplied by BBC.

[b] Neither Lingen nor Gundremmingen have operated since January 1977 and have now both been permanently closed down. Lingen incorporated an oil/gas fired superheater; 160 MW were produced by the nuclear island with the remainder produced by the superheater.

[c] HEW originally owned 50% of Brokdorf but in 1985, 60% of its holding was sold to Preussenelektra.

Source: SPRU nuclear power plant data base.

PWR development has been carried out primarily by Siemens whilst BWRs were developed by AEG. In 1969 these two companies merged their reactor interests to form Kraftwerk Union (KWU) (see Surrey & Chesshire, 1972). Subsequently, in 1976, Siemens bought out AEG's 50% stake in KWU and AEG no longer has any major reactor supply interests. From October 1987, KWU will lose its status as an independent wholly-owned subsidiary of Siemens and become a division of Siemens. KWU has also won orders from Argentina for two heavy water reactors (one completed in 1974 and one under construction) of its own design. Through its wholly-owned subsidiary, Interatom, it is supplying the FBR prototype at Kalkar and it is carrying out research on high-temperature reactors.

The Swiss–German company BBC was initially involved in high-temperature reactor work and, through its subsidiary Hochtemperatur Reaktorbau Gesellschaft (HRB), won the order for the prototype high-temperature reactor (HTR) at Uentrop. Following the amalgamation of AEG and Siemens' reactor interests, the major utilities, especially RWE, encouraged BBC to maintain competition in LWR supply by becoming an alternative supplier. This they did by forming a joint venture with US Babcock & Wilcox (B&W) in 1971, Babcock Brown Boveri Reaktor GmbH (BBR), offering the US B&W design of PWRs. Initially B&W owned 74% of this company but subsequently BBC increased its holding in several stages and now BBR is a wholly-owned subsidiary. BBR won only three orders for reactors, two of which were cancelled and the other of which was recently completed. BBR is no longer actively seeking reactor sales and it seems unlikely that it will build any further units.

KWU does not manufacture any of the major nuclear components but is able to draw on a number of powerful German heavy engineering companies for the supply of components such as pressure vessels and steam generators. In addition, where it has felt it appropriate, KWU has placed orders for such components outside Germany in countries such as Italy, Austria, Switzerland, Sweden, Spain, Japan and the USA. Such orders for components may have been placed on cost and quality grounds and may also have been placed to improve prospects for export sales.

KWU has focussed its efforts on building up expertise in the software side of the business, notably in design and in control systems. Unlike its American competitors, KWU also carries out the architect-engineering for its orders.

Government

As FR Germany is a federal state, government operates at two levels, the federal authorities and the constituent *Land* authorities which exercise a significant amount of power. Government, particularly at the federal level, has consistently tried to place nuclear power in as free a market as possible, minimising subsidies and intervention. In practice, this leaves the federal government with the task of setting the legal and regulatory framework, sponsoring long-term research, development and demonstration (RD&D) of pre-commercial activities such as developing new reactor types, and setting the energy supply and demand framework.

RD&D is controlled by the Ministry for Research and Technology which oversees the national research centres and funds demonstration projects. The Ministry for Economic Affairs commissions supply and demand forecasts and has encouraged nuclear projects without actually offering financial support.

Regulators

The Federal Ministry of the Interior sets the regulatory framework for safety which is administered by the *Länder*. The regulations are based on the German Atomic Law first passed in 1960 and made substantially more stringent in 1976. The Atomic Law (plus the ordinances which the Act empowers the Federal Government to issue) provides a detailed and comprehensive framework within which licensing takes place. The Ministry of the Interior is advised in these matters by two committees, the Commission on Reactor Safety (RSK) and the Commission on Radiation Protection (SSK). These committees include independent experts as well as Ministry staff. The *Land* authorities are advised by the TUVs (Technical Control Services) which are semi-public engineering bodies organised at the *Land* level. They are responsible for certifying the safety of a wide range of equipment besides nuclear power plant, for example, they carry out the annual mandatory boiler and vehicle inspections and tests.

One of the features of reactor construction has been the very thorough testing procedure applied to primary parts, safety valves and emergency cooling systems. These are subject to tests first by the manufacturer, then by KWU and finally by the TUV giving considerable confidence that components are of high quality.

Whilst the rules for licensing are set at a federal level and despite the very comprehensive nature of these rules, there is scope for differing interpretations of these rules by the *Länder*, and the federal authorities cannot readily overrule *Länder* decisions.

Reactor licensing, although nominally a one-licence system, is in practice, a multi-stage process. In the first stage, the applicant makes an application to the relevant *Land*. At this stage the *Land* makes a public announcement and allows two months for public objections to be raised.

This is followed by a public hearing, where the judge has powers to determine whether or not the proposal meets the provisions of the Atomic Law. The same sequence applies to a whole series of further permits that must be sought as construction proceeds. The provisions of the 1976 Act, which requires all plants to embody current best practice technology, mean that at each permit application, backfitting or design changes may be required if any part of the existing construction does not conform to current best practice. Appeals against the decisions of the local administrative courts have, until recently, been made through a three-tier hierarchy of courts, each of which can take three to four years to pass judgement. The three tiers were the *Verwaltungsgericht* (of which there are many), the *Oberverwaltungsgericht* (one for each *Land*) and the *Bundesverwaltungsgericht*, the highest court. The lowest tier now no longer hears appeals on this subject. At the highest level, only questions of legal interpretation are considered rather than substantial design and safety issues.

In theory, any contested permit application leads to an automatic suspension of work, but the government may by-pass this by means of an immediate effect order (IEO), which declares the project to be in the public interest: this over-rides the initial suspension of work. The judge may then rule that the IEO itself be suspended if he or she believes that the objectors have a strong case. This illustrates the strongly judicial character of the regulatory process.

Economic regulation is also carried out at the *Land* level. Large contract prices are regulated by cartel law whilst tariff prices are regulated by the *Land* authorities. These authorities have decided that if they are to judge whether tariff rates are fair, they have to know what prices are being charged in large contracts. Thus, the system of regulation is reasonably comprehensive. The issue of charging customers for plant not yet in service, which has become of considerable importance in the USA, is not a major issue in Germany.

Political influences *

Until recently, at the highest party levels, all the major parties have supported nuclear power. However, at a local level, support is much less even and this has had an important effect on the siting and speed of construction of plant. The right-of-centre Christian Democrats (CDU/CSU) have consistently supported nuclear power, and are strongly critical of anti-nuclear activity. At a *Land* level, this support is reflected in the fact

---

* For a more comprehensive account of the political influences, see Nelkin & Pollak (1981).

that the *Land* with the most stable environment for nuclear power, Bavaria, has always had a CSU government with an absolute majority.

The Social Democrats (SPD) and Free Democrats (FDP) are much more ambivalent. Much of their leadership is committed to continued support for nuclear power. In this they are followed by the greater part of SPD's trade union representation, but opposed by the party's youth organisation and some of the local SPD *Land* governments. In particular, the SPD government of North-Rhine Westphalia, the heart of Germany's coal country, has been at best lukewarm about nuclear power, and only one reactor, Würgassen, has been constructed, despite a population of over 17 million. The Hamm project, ordered more than 10 years ago, has still to receive clearance for construction to start and one of the utilities represented in North-Rhine Westphalia, VEW, has chosen to site its latest unit, Emsland, in Lower Saxony which has a CDU government.

The 'Green' movement has recently become of significant political importance. In the 1982 elections, they won representation in the Bundestag for the first time, and, particularly at a local level, the policies of the Greens are being adopted. This has been a result either of the Greens participation in ruling coalitions (as in Hesse) or the major parties attempting to placate Green supporters without allowing Green representatives into government (as in Saar).

Trade union attitudes towards nuclear power have generally been favourable. The large (9 million members) German trade union federation (DGB) has pursued a policy of support for nuclear power, tempered only by some detailed reservations about the importance of safety and waste disposal. Among constituent unions, the metallurgy and public service unions have made stronger criticism, but solid support has come from the mining, construction and chemical unions.

Opposition to nuclear power has frequently come from groups outside conventional politics. The older environmental associations, with an apolitical tradition, have largely avoided commitment on the nuclear issue: the main focus of protest has been citizen initiatives, which have burgeoned in FR Germany on a variety of issues since the late 1960s.

In 1972, an umbrella organisation (BBU) was formed by 15 local groups, which, despite other interests in pollution and transport problems, by then had opposition to nuclear power as its chief focus. By 1977 it could claim 950 affiliated groups but, from about 1980 onwards, its focus moved more towards opposition to nuclear weapons. Despite a fairly wide political spectrum, the core of BBU support comes from the left. The organisational style of BBU is that of a social network rather than a formal organisation, but, in addition to supporting legal challenges to nuclear projects, it has been able to mobilise large numbers of people (often in tens of thousands) to attend anti-nuclear demonstrations. Its engagement in the peace move-ment has led to a number of internal disagreements and this, coupled with the loss of leading personalities, has severely reduced its effectiveness and influence.

THE PRE-COMMERCIAL ERA (TO 1962)*

Although war-time research had produced some expertise in heavy water and graphite moderated reactors, substantial development of civil power reactors did not begin until the mid-1950s. The absence of any military content to West Germany's post-war nuclear research has meant that there has never been a need for a government controlled body such as the British UKAEA, the French CEA or the American AEC which had wide influence over the directions of technology development and considerable political power. Nevertheless fears of serious fuel shortages arose in the mid-1950s as indigenous coal output declined resulting in heavy dependence on imported fuel supplies. Together with the availability of strategic material under the McMahon Act, this led to the establishment of the German Atomic Commission in 1956. The commission was set up to advise the newly created position of Atomic Minister on matters such as legislation, research priorities and training, a position it held until its dissolution in 1971.

Also in 1956 the research centres at Karlsruhe and Jülich were set up and these have remained the main focus of publicly-funded nuclear research. Jülich has worked mainly on high temperature reactors, whilst Karlsruhe, after early heavy water reactor work, has been the principal centre of fast breeder activity. Six other federally-funded research establishments were subsequently set up but are funded at a much lower level and are not so centrally involved in reactor development.

On the industrial side, a number of companies became involved, notably Siemens and AEG, which were able to use historic links to license LWR technology from Westinghouse and General Electric respectively although at that time Siemens was mainly interested in heavy water reactors. Brown Boveri and Krupp led the development work on high temperature reactors, and Deutsche Babcock developed the PWR for civil marine propulsion. These industrial interests were instrumental in setting up the German Atomic Forum in 1959 which subsequently became a powerful sponsor of nuclear power.

The Atomic Commission, through a subcommittee dominated by heavy electrical and chemical manufacturers and academics formulated the

---

* The material for the historical sections is partly drawn from Keck (1981) and Winnacker & Wirtz (1975).

'Eltville programme' in 1957. This envisaged that 5100 MW of reactors of German design would be in operation by 1965 with an initial 500 MW programme which was to be promoted using state funds. Foreign reactor types such as those under development in the UK and USA were to be supported in Germany only if they did not conflict with the development of German reactor types. Underpinning this programme was a belief that no existing design of reactor was ideal and that current reactor developments were only interim stages on the way to the breeder reactor which was regarded as the only viable reactor type in the long-term. The Eltville programme was re-evaluated in 1958 and five reactor concepts emerged as being most worthy of development. These were:

– a pressurised heavy water cooled, heavy water moderated natural uranium reactor
– a light water moderated gas-cooled reactor
– a gas-cooled, graphite moderated, natural uranium reactor
– a high-temperature reactor with gas-cooling and enriched uranium
– an advanced reactor cooled by an organic fluid

Provision for super-heating and recycling of plutonium reflected the concern to maximise the utilisation of the ultimately scarce resource, uranium. It was assumed that the programme would be financed by utilities, with some tax reductions and government loans.

Whilst some of these concepts have proved the basis for subsequent development, little solid in terms of reactor construction emerged from the Eltville programme and the subsequent (1959–60) 'advanced reactor programme'. The utilities had been poorly represented in this exercise and had a strong preference (unlike the government, manufacturers or the academics) for assessing nuclear power on the basis of international designs, with international tendering necessary if orders were to be placed. They also felt that a size of 100 MW for prototype plants involved an excessive risk and they favoured plants of around 15 MW.

Reactor construction took place primarily outside the Eltville programme under the auspices of Karlsruhe, Jülich and the utilities, particularly RWE. At Karlsruhe, the FR2 heavy water reactor, the first German reactor, was completed in 1960. The reactor produced 12 MW(th) and, being for research purposes only, produced no electricity. Commissioning of this reactor was followed immediately by the start of work on the 50 MW(e) MZFR 'multi-purpose' reactor supplied by Siemens and also situated at Karlsruhe. Although this was nominally a research reactor, much of its expected function of materials testing was carried out in the FR2 reactor and for most of its life it was used largely as a power reactor. MZFR was completed in 1965 but confidence in the concept was such that Argentina agreed to purchase a commercial-sized unit, Atucha 1

(340 MW), from Siemens in 1962 although the order was not formally placed until 1968. A high temperature reactor based on the 'pebble-bed' design* was ordered from BBC/Krupp in 1959. This unit sited at Jülich produces 15 MW(e) and was finally commissioned in 1969 after a lengthy construction and commissioning process.

Perhaps the most significant order, however, was the agreement between RWE and Bayernwerk with General Electric (USA) for the supply of a 15 MW(e) BWR (VAK) to be sited in Kahl. The NSSS was supplied by General Electric with AEG as the main contractor and architect-engineer. The reactor went critical in November 1960.

All four of these prototype units have proved successful and are only now coming to the end of their lives, considerably outlasting some of their successors. In 1962 RWE and Bayernwerk, encouraged by the success of the VAK reactor, placed on order for the first commercial-size reactor in Germany, the 250 MW Gundremmingen BWR. This order came within the framework of the Euratom–United States power station programme. As with the VAK order, the NSSS was supplied by General Electric with AEG as its German collaborator. Despite this order's contravention of the spirit of the Eltville and advanced reactor programme agreements, the federal government was forced to the conclusion that the LWR route was the quickest way of establishing a German reactor industry and they provided subsidies and guarantees against operating losses.

Siemens, who had obtained a PWR licence from Westinghouse in 1955, had started work on a design for a commercial-sized PWR after the completion of the MZFR design, but were not yet in a position to make a bid for the Gundremmingen order. Siemens favoured the PWR over their heavy water design because the heavy water units required a much larger pressure vessel than a PWR of similar electrical output and seemed to restrict development to units not much larger than about 350 MW.

In some respects the Gundremmingen order parallels the 1963 Oyster Creek order in the USA, concentrating development on the LWR and its suppliers. However, unlike the USA, reactor development both of LWRs and other concepts did continue, particularly up till the end of the next phase of nuclear plant development (1969) and government subsidies on reactor construction were also continued for some time.

THE COMMERCIAL DEVELOPMENT ERA (1963–69)

During this period, both Siemens and AEG sought to develop and improve their LWRs over their American parent designs. AEG's attempts were

* In the pebble-bed reactor, the fuel is contained in a large number of ball-type fuel elements which are arranged to allow simple on-line refuelling.

more radical and reflected the long-standing concern that LWRs were not as thermally-efficient as fossil-fired stations due to the lower temperatures and pressures of the steam produced by LWRs. This problem was to be solved by use of a superheater which brought the steam up to the same temperature as that produced by a conventional plant. A 25 MW prototype plant, Grosswelzheim, which used nuclear energy to superheat the steam was ordered in 1963 whilst a less radical but commercial-size design using oil (subsequently gas) to superheat the steam was ordered in 1964 for installation at Lingen. The Lingen unit produced 160 MW from the nuclear island boosted by a further 90 MW from the superheater. Even before either of these units was completed, AEG had abandoned the idea of superheating and for their next order, Wurgassen in 1967, they offered a design very similar to the General Electric Mark 2 BWR, which had also abandoned the use of secondary steam generators in favour of the simpler direct cycle. The Lingen unit was completed in 1968 and operated reasonably well for 10 years before it was forced, like its predecessor Gundremmingen, to shut down permanently following the discovery of serious pipe-cracking (see later). Grosswelzheim was not completed until 1970 and a combination of operating difficulties and lack of interest in the concept meant that it was permanently closed only a year later.

Whereas AEG's early design work can be characterised as trying to improve the BWR concept, Siemen's design work was much more aimed at optimising the existing concept by improving materials, manufacturing methods and water chemistry. In many ways, the prototype for the PWRs was the MZFR PHWR which also uses a large pressure vessel. Siemens had held a licence from Westinghouse since 1955 and when they won their first order for a commercial-size PWR in 1964 (Obrigheim, 345 MW), Siemens had already made significant changes to the construction of the reactor pressure vessel. Obrigheim was the last of the three demonstration LWRs which received direct government subsidy, all subsequent orders being wholly utility financed. Two further orders were placed for LWRs prior to 1969 both of which can be seen as transitional in size and design. The Wurgassen BWR order marked the return by AEG to a design very close to the GE parent design. The Stade PWR order incorporated some important improvements to the materials used in the steam generator but had not adopted the modular loop concept that all subsequent PWRs embodied. Stade was a 622 MW 4-loop design, whilst for all subsequent orders KWU used designs with three or four loops, each loop having a capacity of about 300 MW. These were the last commercial orders to be placed before the merger of Siemens' and AEG's power station businesses in April 1969.

Little in the way of reactor construction resulted from the government's second nuclear programme which ran from 1963–67. The 100 MW Niederaichbach reactor built by Siemens, a carry-over from the Eltville programme, was ordered in 1964. This design had a number of theoretical attractions; the use of carbon dioxide as coolant allowed higher coolant temperatures and thus better thermal efficiency; the use of heavy water as moderator allowed the use of natural or only slightly enriched uranium; and the use of pressure tubes similar to CANDU pressure tubes avoided the need for a large pressure vessel. However, the success of the LWRs and extreme difficulties encountered during Niederaichbach's commissioning (such that two years after first criticality in 1972, it had still only achieved 40% of design rating) meant that the plant was permanently shut down in July 1974.

Fast reactor work, which had started at Karlsruhe in 1960, proceeded via an order in 1966 for an experimental unit, KNK, of 20 MW which was planned to be the first step towards the building of a 300 MW prototype FBR.

TECHNOLOGICAL MATURITY (1969–75)

In many respects, 1969 seemed to mark the maturity of the German LWR industry. Siemens and AEG merged their power plant interests in the jointly owned Kraftwerk Union. However, although all reactor sales were placed through KWU, because of restrictions in the American technology licences, the reactor divisions operated separately with the Siemens PWR team at Erlangen and the AEG BWR team at Frankfurt and were not transferred to KWU until 1973. Shortly after the merger, KWU won its first LWR export order for a PWR for the Netherlands (see Table 6.6). It also won an order for its first unit larger than 1000 MW (the Biblis 'A' PWR) and its first order for a large BWR, the 800 MW Brunsbuttel unit, the first of six 'BWR–69' reactors, one of which was an export to Austria.

In 1970 Siemens confirmed its independence from Westinghouse by terminating its licence agreement some six years earlier than scheduled. Contacts with Westinghouse had been scarce during the last years of the agreement and, in 1971, there was some prospect of Westinghouse competing in Germany with Siemens for reactor sales. Technologically, the PWR side of KWU continued smoothly through this period with excellent operating performance and short construction times. Fundamental design changes, particularly after the ordering of the second unit at Biblis in 1972 were much fewer and negotiations for export markets to Switzerland, Spain, Iran and Brazil were progressing well.

The AEG side of KWU was less successful. The Wurgassen unit went critical in 1972 after a construction period of only 46 months. However, a series of technical problems (see later for a fuller account), some of which had implications for the subsequent BWR–69s, meant that it was not until

Table 6.6. *German nuclear exports*

Country	Name	Type	MW (gross)	Year of order	Year of commissioning	NSSS supplier	Turbine generator supplier
Argentina	Atucha 1	PHWR	357	1968	1974	Siemens	KWU
Netherlands	Borssele	PWR	477	1969	1973	KWU	Smith/Stork/Siemens
Austria	Tullnerfeld	BWR	722	1971	–[a]	AEG/KWU	KWU/Elin
Switzerland	Gosgen	PWR	970	1972	1979	KWU	KWU/Ensa
Spain	Trillo 1	PWR	1040	1976	1990	KWU	KWU/Ensa
Brazil	Angra dos Reis 2, 3	PWR	2 × 1340	1976	1994	KWU	KWU
Iran	Bushehr 1, 2	PWR	2 × 1293	1974	–[b]	KWU	KWU
Argentina	Atucha 2	PHWR	745	1981	1994	KWU	KWU
Spain	Trillo 2	PWR	1040	1981	–[c]	KWU	KWU

[a] The unit was completed in 1979, but in a referendum, it was voted that the unit should not be allowed to start up. The unit will now probably be dismantled and sold.
[b] Construction work was suspended on these units following the Iranian revolution in 1979. Work on unit 1 was nearly 90% completed and work on unit 2 was about 60% complete. There are no firm plans to complete these units.
[c] Work on this unit was suspended following the 1983 Spanish general election. There are no current plans to complete it.
Source: SPRU nuclear power plant data base.

October 1975 that the utility, Preussenelektra, finally accepted handover of the unit from the vendor (even then output was limited to 80% of design). These problems caused AEG heavy losses which made it difficult to meet the capital demands of KWU. After three years of speculation about the fate of AEG's stake in KWU – numerous alternative partners were mooted – Siemens bought out the entire AEG stake on 1 January 1977.

Particularly in the early part of this period, BBC's role in reactor supply was not settled. There were suggestions that it might become a supplier of British designed AGRs and there was uncertainty about the role of its partner Krupp in their joint venture company, Brown Boveri/Krupp Reaktorbau (BBK), which developed high-temperature reactors. In 1971 the situation was clarified; BBC reached agreement with B&W, USA to set up the joint venture BBR to supply the American company's design of PWR. Initially 74% of the holding was held by B&W but this was reduced to 40% in 1978 and in 1981 BBC bought out the remainder of the holding. BBC was encouraged to supply PWRs by the utilities, particularly RWE, which were apprehensive about the monopoly position KWU then held in LWR supply and RWE provided BBR with its first order in 1973 for the Mulheim Kaerlich unit – this is likely to be BBR's only completed order as two subsequent orders were not built and BBR is no longer actively seeking orders.

Krupp withdrew from BBK in 1971 and a new company was formed in 1973, HRB, in which Krupp was replaced by the US supplier of high-temperature reactors (of a somewhat different design concept), Gulf General Atomic. Shortly after Krupp's withdrawal, a consortium of utilities with Euratom and federal government financial support placed an order with HRB for the 300 MW Uentrop HTR which, like its predecessor AVR, was based on the pebble-bed principle.

Soon after this, in 1972, an order was placed for the 300 MW Kalkar FBR. This represented the other (and larger) strand in the third and fourth nuclear programmes (1968–72, 1973–76) of federal funding for reactor development. The reactor was to be supplied by an international consortium involving German, Dutch and Belgian companies with the 70% German share being taken by Interatom, which had supplied the KNK experimental FBR and in which Siemens had recently taken 100% ownership buying the remaining 25% of shares from Deutsche Babcock.

The federal programmes also devoted a higher proportion of resources to fuel cycle activities than previously, although at this stage most money was spent on the 'front-end' of the fuel cycle. The Urenco venture was set up in collaboration with the Netherlands and the UK to build gas ultra-centrifuge enrichment plants. The government's contribution to this project was planned to be the second largest discrete expenditure in the fourth nuclear programme after the breeder but ahead of the high-temperature reactor. The more controversial 'back-end' of the fuel cycle was given less emphasis with reprocessing of LWR fuel to be carried out at La Hague in France and Windscale in the UK. Only when these facilities were fully exploited was a plant to be built in Germany.

After the Brunsbuttel and Biblis A orders, about three units a year were ordered from 1970 to the end of 1974. However, opposition to these units was strong and the timelag between ordering and start of major work on site grew considerably. In addition, safety requirements escalated, including the increased requirements to retrofit safety systems both to reactors under construction and to those already in operation.

Nevertheless, the oil crisis seemed to underline the need for nuclear power and, in 1974, the federal energy programme foresaw a need for 20 GW of nuclear power to be in operation by 1980 rising to 45–60 GW by 1985 when 45% of electricity would be nuclear generated. This projection severely overestimated nuclear capacity and electricity demand growth. In fact nuclear capacity had not reached 15 GW by 1985 and 50 GW of nuclear plant would have produced almost all of Germany's requirements for electricity.

THE PEAK OF OPPOSITION (1975–79)*

Whereas in France, direct government links with EdF meant that the projected nuclear power needs could be turned directly into plant orders, in FR Germany the forecasts were indicative and the federal government could only indirectly influence the ordering of nuclear plant. Nevertheless, a change in German investment legislation allowed a burst of ordering for nuclear plant in 1975 when the ordering of ten units seemed to confirm the direction of policy.

A number of these sites, plus the proposed breeder reactor at Kalkar and the major fuel cycle facility to be built at Gorleben in Lower Saxony, including reprocessing and waste disposal announced by Chancellor Schmidt in 1976, proved the focus for concerted opposition. The most spectacular events occurred in 1976 and 1977, when site occupations and serious violence took place. The first major confrontation was at Wyhl, where, after almost a year of site occupation, demonstrators agreed to leave in early 1976 in return for a promise that their case would be more seriously heard in court actions. In late 1976, police drove off protesters occupying Brokdorf, to the accompaniment of some violence. The worst battle of all, with hundreds of injuries, took place in 1977 in Grohnde. An attempted

* See Nelkin & Pollak (1981) for a much fuller account of this period.

occupation at Kalkar was prevented by major police action, and although there were some disturbances at Gorleben in 1979, direct confrontation was essentially over by the end of 1977.

While confrontation made political headlines, it would, in the face of growing and eventually massive police resistance, have achieved no more than a few relatively minor construction delays had not the administrative courts simultaneously started to uphold some of the protesters' views. The two key events in this respect were the 1972 decision of the supreme administrative court that safety considerations were always to take precedence over economic arguments in the licensing process, and the 1976 revision to the Atomic Law, which, most importantly, insisted that all plants must embody best-practice technology. Armed with such guidelines, judges who saw themselves as defenders of individuals' rights and who felt sceptical about the claims of the utilities could, and frequently did, reach judgements that severely disrupted utility plans. Examples are the 1977 decision to suspend construction at Wyhl because of apparent inadequacies in the concrete containment, and the withdrawal of the Brokdorf construction licence in the same year because of inadequate waste disposal planning. In the first case, the 1976 best-practice technology ruling provided the main argument, while at Brokdorf it was the priority of safety over economics.

Opposition was the most visible feature of this period. But a number of other developments of long-term importance were also taking place. The publication in 1975 of the US Rasmussen report, which estimated the probabilities of various classes of reactor accident, was followed up in Germany by the commissioning by the Federal Ministry for Research and Technology of the German Risk Study in 1976. The study was published and adopted by the RSK in 1979 and its findings became a legal requirement. In practical terms, probably the most important impact was the adoption of the Basis Safety Concept which had been under development since 1972. The Basis Safety Concept was designed to render the probabilistic approach of the Rasmussen report unnecessary by designing the plant such that catastrophic failures could be regarded as incredible. This meant that a number of mooted safety measures such as burst protection for the reactor pressure vessel need not be incorporated because the failure they were designed to protect against was deemed incredible. The Basis Safety Concept was thus an important factor in the slowing of the proliferation of safety measures.

In terms of reactor performance, whilst the PWRs were operating fairly smoothly, the BWRs were experiencing severe difficulties due to inter-granular stress corrosion cracking (IGSCC). This issue is dealt with in greater detail later, but the discovery of this cracking and the new requirements resulting from the adoption of the Basis Safety Concept led to the two oldest reactors being shut down permanently and the four subsequent BWRs undergoing shutdowns of about a year for major repairs. This effectively removed the BWR, at least temporarily, from the market as a feasible option. However, although replacing all primary pipe-work was a radical solution compared to most countries' approaches, it did at least give the promise of an end to the problem once the repairs were completed.

By 1979, the German system of building and operating nuclear plant appeared to have reached crisis point with most projects apparently thoroughly enmeshed in legal disputes, little prospect of new orders getting off the ground and export markets giving little benefit.

THE RECOVERY PERIOD (1980 ONWARDS)

Despite the apparently bleak outlook for the nuclear industry, a number of factors were beginning to move in its favour. The adoption of the Basis Safety Concept slowed the pace of safety-related design changes. The method of operation of KWU, being a design and software supplier rather than having extensive hardware supply facilities meant that, unlike for example Framatome, KWU was much more flexible in terms of its order requirements. It had never built up an infrastructure of dedicated hardware suppliers fed on a steady diet of orders. Ironically, the massive amount of research and validation work necessary to win judgements in the courts meant that design teams were kept busy and skills were not dissipated. The 'convoy' system (described below), designed to overcome the inconsistency of requirements between different *Länder* and to increase standardisation, had been under development since 1976 and was beginning to gain wider acceptance.

Public opposition seemed to have peaked and court judgements were beginning to go in the industry's favour. Against these factors, the failure to make adequate arrangements for the back-end of the fuel cycle – waste disposal and reprocessing – was beginning to cause embarrassment.

The roots of the convoy system go back to the early 1970s when the industry began to look to standardisation as a means of controlling costs and simplifying licensing. The Biblis B and Unterweser units and, to a lesser extent, Grafenrheinfeld were conceived as essentially identical re-actors. In 1974 the major utilities and KWU came together to form a num-ber of working groups aimed at defining the specification for a standardised PWR. By 1976 the decision was taken to apply these studies and licensing procedures were launched for three units, Biblis C, Vahnum and Hamm. In fact, due to licensing difficulties, no substantial work has been completed on any of these units, some 10 years after this decision. Nevertheless, development of the concept was continued and in 1981 the first convoy was set in motion.

The main features of the convoy system are:
- one common set of engineering documents applied to all projects
- only four licensing steps, three for construction and one for commissioning in contrast to the large number of steps (up to 15) that had applied to previous projects

The four licensing steps cover:
- the documentation necessary for the building construction (about 15 months before start of construction)
- the documents related to mechanical and electrical components, and systems fabrication and erection (about 10 months before start of construction)
- the remaining construction activities (about five months before start of construction)
- commissioning

It is hoped that by use of identical documents complying with the new detailed requirements of the RSK and the Nuclear Standards Safety Committee (KTA), the expert appraisals of the different TUVs can be co-ordinated and agreed leaving only the checking of site-specific details to the local authorities.

This procedure requires a great deal more early design work than previously so that the design in detail is complete before start of construction. It also relies on the compliance of the TUVs and the *Länder* involved.

Optimistically, it was hoped that seven units would form the convoy including four ordered in 1975 which had not been licensed (Wyhl, Biblis C, Hamm and Neckarwestheim 2) but in fact only the last of these and two other plants (Isar 2 and Emsland) were included. At the time of writing, despite a number of legal difficulties, the convoy is progressing reasonably well.

On the operating side, there has also been considerable improvement (covered in greater detail later). The latest PWRs to enter service have operated outstandingly well and major overhauls to the earlier units in 1979–80 seem to have been effective in improving their performance. The earlier BWRs have also been thoroughly overhauled during the lengthy shutdowns for pipe replacement and these are also now showing signs of having overcome the severe quality problems that affected their early years of operation. The newest generation of BWRs whose design was completed after the takeover of KWU by Siemens have performed on a par with the newest PWRs and it may soon be felt appropriate to relaunch this design.

Other activities have not been so uniformly successful. In common with all other reactor vendors, KWU has found export orders difficult to win. It has failed to sell reactors in, perhaps, the most substantial market, South East Asia, and its South American customers in Brazil and Argentina are finding it difficult to finance existing orders and are most unlikely to place further orders. KWU has competed hard for orders in Turkey, Egypt and China but arranging finance for these orders has proved difficult for all vendors and KWU has not yet won any orders in these markets.

Fuel cycle development has also suffered a number of set-backs. Lower Saxony vetoed the construction of a reprocessing plant at Gorleben which would have processed 1400 tonnes of fuel per year and more modest plans are now being formed to construct a number of reprocessing plants of a quarter of this size in different *Länder*. Some progress has been made in finding and getting approval for sites – the first site chosen is Wackersdorf in Bavaria with KWU as the main contractor – but opposition remains strong, as it does for the development of high-level waste disposal sites.

Of the prototype reactors under development, the high-temperature reactor at Uentrop has fared much better than the Kalkar breeder. Uentrop is now complete and FR Germany may be well placed to profit should the high-temperature reactor emerge as a viable technology in the USA where there is renewed interest in this concept. The Kalkar breeder by contrast has suffered from public opposition and the international loss of interest in this reactor type. This has resulted in funding difficulties with its international partners as costs escalate. Although the unit is now substantially complete, it continues to face opposition in the courts to its commissioning and from the *Land* government of Northrhine-Westphalia.

The resumption of public demonstrations of opposition to nuclear developments, in 1986, particularly to the Wackersdorf development suggests that public opposition will remain a strong feature for the foreseeable future.

ECONOMIC PERFORMANCE

Operating performance of German PWRs

By far the most important factor in explaining the performance of German PWRs has been technology maturation (the improvement of all units as overall experience increases) rather than unit maturation (the improvement of each unit through time as construction and design faults are corrected). Thus, the first PWR of modern design, Biblis A, has had, by German standards, a disappointing operating record (nevertheless exceeding the performance of any US PWR of comparable size). By contrast, the units completed since 1979 have had outstanding performance from the outset with virtually no teething problems – typically they have had capacity factors in excess of 85% in their first calendar year of operation. In

Table 6.7. *Performance of German PWRs by time period*

	All reactor years		Mature units only	
	Capacity factor	No. of reactor years	Capacity factor	No. of reactor years
1976–78	64.0	7	70.7	1
1979–80	60.4	7	56.3	5
1981–85	80.8	23	79.9	20
1986	75.7	7	68.9	5

Source: SPRU nuclear power plant data base.

view of this, and the apparent lack of any age-related processes such as steam generator leaks, the detailed analyses of performance (see Table 6.8 and Table 6.9) are split up according to calendar year and the first two years of reactor operation are not omitted.* The analyses cover three time periods (pre-1979, 1979–80 and post 1980. During 1979 and 1980 three of the units underwent major maintenance, refuelling and repair shutdowns taking four to six months, and the two Biblis units' output was severely restricted by a regulatory problem (discussed in detail later).

PWR performance is summarised in Table 6.7 and analysed in greater detail in Table 6.8 and Table 6.9.† Table 6.7 shows that omitting the early performance of units does not consistently increase the mean capacity factors suggesting that whilst there is likely to be unit maturation, it is much less important than other factors. It does illustrate the relatively poor early performance of the Biblis units in 1976–78, the lengthy shutdowns during 1979 and 1980 and the excellent performance since then. The causes of lost output are analysed in Table 6.8,‡ with the high operating losses in 1979

---

* For the reactors in the other three case study countries, performance in the first two years is significantly affected by the remedying of design and construction errors and general 'teething' problems. This means that the performance in these two years is not typical of later performance and thus, is omitted from some analyses.

† The first two units Obrigheim (345 MW) and Stade (662 MW) are omitted from the analyses on the grounds that their designs differ significantly from those of current units. However, two factors are worth noting about these units. First their long-term performance has been outstanding with lifetime capacity factors in excess of 80%. Second, unlike its successors which used Incoloy 800, the steam generators for Obrigheim used the same material as Westinghouse, Inconel 600. These have long been regarded as unsatisfactory, requiring regular repairs due to corrosion and, in 1983, they were replaced with new units made of Incoloy 800. Whereas in the USA such repairs have taken up to 15 months, factors such as the large amount of working space in the containment restricted the shutdown time to only 3 months, during which time the unit was also refuelled.

‡ The analyses presented in Tables 6.8 and 6.9 are similar to those shown in the appendix of Chapter 5 covering US units. The significance of such analyses are discussed in that appendix.

Table 6.8. *Causes of lost output in German PWRs according to time period (%)*

Time period	No. of reaction years	Capacity factor	Shutdown losses	Derating losses	Operating losses	Adjusted operating losses
1976–78	7	64.0	27.6	2.8	5.6	7.7
1979–80	7	60.4	26.9	1.4	11.3	15.6
1981–84	18	80.0	16.6	0.1	3.2	3.9

Source: SPRU nuclear power plant data base.

and 1980 the result of the regulatory restriction at the two Biblis units. The causes of shutdown are analysed in Table 6.9, showing the extremely good reliability of the various plant systems with few breakdowns and no consistently troublesome areas. The regulatory authorities did require a number of quite substantial testing and inspection shutdowns in the earliest period. Often these related to systems which had suffered a significant failure. For example in 1976, following the discovery of defective bolts in the feedwater system and cracks in the feedwater tank at Biblis A, regulatory action extended the 3-month shutdown by a further month for inspections to be carried out.

The regulatory restriction at the Biblis units arose from inadequate provision for spent fuel. In normal operation, one-third of the fuel for a PWR is replaced at each refuelling and, by German law, the spent fuel store must be able to hold five-thirds of a reactor core, and there must be sufficient storage space for the reactor core to be completely unloaded. With delays that have occurred in fuel cycle service developments, particularly the abandonment of the proposed Gorleben facility and with La Hague unable to accept spent fuel until 1981, the utility, RWE, expanded on-site spent fuel storage to accommodate several years of fuel. However, in contrast to US practice, spent fuel storage is sited within the containment for safety reasons.

It was decided that the addition of further storage constituted a modification requiring full-scale licensing procedures, including public hearings. A massive number of public objections were registered and the case took more than four years to resolve. The basic point of contention was whether the effects of a collision of two tankers in the nearby Rhine could be withstood by the containment within which the spent fuel was stored. This illustrates the way in which minor modifications to reactors can re-open questions of safety that have apparently been resolved. The net result of this problem was that both Biblis A and B operated at only 50% capacity for 18 months in order to keep the volume of spent fuel in storage within the legal limits, pending the resolution of the case. Finally the problem was alleviated when La Hague began to accept shipments of spent fuel.

Table 6.9. *Causes of shutdown in German PWRs according to time period*

Time period	No. of reactor years	Shutdowns[a] (Hours per year of operation)											Shutdown frequency (no. per year of operation)
		4	5	6	7	8	9	C	D	E	G	H	
1976–78	7	14	110	19	83	208	0	1566	565	306	153	0	8.3
1979–80	7	0	0	0	0	9	0	2206	688	141	0	0	3.9
1981–84	18	0	24	119	8	240	3	1146	45	0	0	2	3.2

[a] Causes of shutdown are:

4 = nuclear auxiliary and emergency system
5 = main heat removal system
6 = steam generators
7 = feedwater condenser and circulating water systems
8 = turbine generator
9 = electrical power and supply system

C = annual refuelling and maintenance
D = maintenance and repair
E = testing of plant systems/components
G = regulatory limitation
H = other

[b] Note that no shutdowns were reported caused by failures in the reactor and accessories, fuel, reactor control system and instrumentation or miscellaneous equipment. No shutdowns were caused by operating error or training and licensing.

Source: SPRU nuclear power plant data base.

Experience with steam generators has been in marked contrast to that of Westinghouse units despite the fact that the basic design concept is very similar. There have been six major corrosion mechanisms affecting Westinghouse steam generators and these are described in detail in the analysis of US Westinghouse units. Denting, crevice corrosion and pitting have been entirely avoided by use of better materials, good manufacturing procedure and careful water treatment. Stress corrosion cracking was found in the first German PWR, Obrigheim, which used Inconel 600 steam generator pipes, after only two years of service. Subsequent units have used Incoloy 800 which has not proved prone to this problem. Wastage, which results from the use of phosphates to control the water chemistry, is a more complex issue. All German units have, until recently, used phosphate treatment but at concentrations of less than a tenth of that typically used in US plant. This has radically reduced the wastage problem and only two tube leakages have occurred. However, some tube plugging has been necessary where the tubes have become too thin. Units entering service now all use all volatile treatment (AVT) from the start. The sixth factor, vibration, is by its nature less tangible, but the good performance of steam generators suggests that it is not significant in German PWRs.

In addition to these differences in the steam generators, KWU is now using corrosion-resistant condenser tubes thus avoiding the presence of corrosive salts on the secondary side of the steam generator.

Operating performance of German BWRs

The operating record of German BWRs is dominated by two factors; the discovery and remedying of intergranular stress corrosion cracking (IGSCC) and the poor early performance of the Wurgassen and Brunsbuttel units.

IGSCC. IGSCC is a problem that has affected BWRs worldwide, being first discovered in the USA in 1974. (See Chapter 5 for a more detailed account of the causes of IGSCC.) Its main impact in Germany however was not till 1977. In January of that year both of the two early units, Gundremmingen and Lingen, experienced forced shutdowns, in the former case due to a fault in a control system and in the latter due to the discovery of cracks in the steam generators (these two units were the only commercial German BWRs to employ steam generators). Subsequent inspection of the plants revealed serious IGSCC in the feedwater pipes, especially at the Gundremmingen plant. It became apparent that all BWRs which used the same type of steel (unstabilised 304 stainless) including those under construction could be affected and did not meet the Basis Safety Concept. For the two small plants, whilst repairs were feasible, the regulatory authorities demanded that additional safety systems, such as a high-pressure Emergency Core Cooling System and an emergency control room, be retrofitted. Even with these extensive backfits, the regulatory authorities could not guarantee long-term licensing. The utilities took the view that the value of these rather small plants was not sufficient to take the risk of further investment in them and both units have been permanently closed down. Their performance up to 1977 had been reasonably good although the steam generators had proved consistently troublesome.

For the four larger units either in operation (Brunsbuttel and Wurgassen) or nearing completion (Isar 1 and Philippsburg 1) at that time, the decision was taken to replace all primary pipe-work with a manganese-nickel alloy steel of lower carbon content. These modifications required a shutdown of about 15 months for each plant and were carried out between May 1980 and October 1983. The three later BWR orders, Krummel,* Gundremmingen B and Gundremmingen C which were declared commercial in 1984 and 1985 were built with the appropriate grade of piping and will not require modification.

Wurgassen. This unit went critical in January 1972 after less than four years of construction work but a series of technical problems meant that it was not declared commercial until October 1975 and then at only 80% of design rating. This delay resulted from three separate problems. First, and most seriously, in 1972, after minor problems with the steam separators, a relief valve stuck in the open position, causing a pressure pulse in the pressure suppression condenser chambers. This resulted in structural damage to the building, and repairs took seven months. Second, in 1973, serious cracking was discovered in one of the four main steam lines at a t-junction and this led to a further five-month shutdown. Finally, in 1974, cracks in the low pressure shaft of the turbine generator caused a fourteen-months shutdown for repairs, and it was not until further problems with the steam separators were resolved that Preussenelektra finally accepted handover from AEG. The safety authorities limited output to 80% of design rating pending modifications to bring the unit up to current safety standards, such as the fitting of an independent residual heat removal system – its original system was connected to the emergency core cooling system. Allowing for the 20% derating, its performance over the next six years was reasonably good, until in 1982/83 the pipe replacement and safety retrofits were carried out, being completed in October 1983. In 1984/85 with the output limitation lifted, it achieved a capacity factor of over 80%.

---

* The Krummel reactor is based on the BWR–69 design but construction was at a sufficiently early stage that the modifications could be readily carried out.

Table 6.10. *Recent capacity factors of German BWRs*

Unit	Commercial operation	Pipe replacement	Capacity factors				
			1982	1983	1984	1985	1986
Wurgassen	11/75	5/82–9/83			79.7	82.4	86.0
Brunsbuttel	2/77	7/82–8/83			78.9	83.3	83.4
Isar 1	3/79	9/81–9/82		93.9	73.1	85.7	83.8
Philippsburg	2/80	5/80–10/81	66.4	72.7	83.3	81.1	69.1
Krummel	3/84	–				84.2	85.6
Gundremmingen B	7/84	–				84.1	76.4
Gundremmingen C	1/85	–				83.7	73.6

Source: SPRU nuclear power plant data base.

The problem with the relief valve resulted in the redesign and retrofitting of the pressure suppression chambers in the other three BWR–69s causing considerable delay to their completion.

Brunsbuttel. The second BWR–69, at Brunsbuttel, also experienced serious technical problems which had major financial consequences. After a relatively trouble-free period of seven months between criticality in June 1976 and commercial operation in February 1977, a pump failure in September 1977 caused a four-month shutdown. Then in June 1978 a pipe leak in the turbine section released radioactive steam into the machinery building. The fault was compounded by a misinterpretation by the operators, who overrode automatic shut-off devices and allowed this steam to escape for three hours. Although in terms of hardware damage the incident was relatively minor, massive changes to the man-machine interface were required, and training for operators was improved. In all, after protesters took up the action in court, the shutdown lasted more than two years. Like Wurgassen, the pipe replacement was carried out in 1982/83 and in 1984/85 the unit also achieved a capacity factor of over 80%.

Future BWR performance. Overall, if the objective is to gain insights into likely future performance, the only relevant experience is that which has been accumulated since pipe-work replacement with the overhauled BWR–69s and the new units – the last BWR–69, Krummel and the two BWR–72s, Gundremmingen B&C. Table 6.10 shows the dramatic and sustained improvement in performance with the older BWRs since pipe-work replacement. The first-year performance of the new BWRs suggests that, like the latest PWRs, these units are very well constructed with few installation errors. If these levels of performance are sustained and the new piping material proves resistant to cracking, there will be a strong case for ordering further BWRs.

Construction times and capital costs

In the analysis of construction times, the three early units below 600 MW – Gundremmingen, Lingen and Obrigheim – are excluded because they were all built with a level of subsidy from the government and are technically unlike the later and much larger plants. Table 6.11 shows the construction time·experience for all completed commercial reactors. There are three reactors currently under construction, all expected to be commercial by 1989.

Table 6.11 is organised by date of construction start, because the clearest statistical trend is towards longer construction periods with later construction starts: a difference on average of 33 months longer construction time for reactors started in 1972 or later compared to those under way by 1970.

This time trend does not in itself provide any explanation for increasing construction times, and in addition there is considerable dispersion around the average times in each of the two periods. What is clear is that the minimum construction time has lengthened considerably. While three of the earlier plants were completed in around five years, none of the later group has taken less than seven years and five months though reactors still under construction are showing signs of slightly faster times than those recently completed. The major generic explanation for these increases in the later period is the effects, as mediated through the courts, of the stringent Atomic Law revisions of 1976, which mandated the use of best practice technology. In general, plants which were complete (or substantially so) by the later 1970s avoided the retrofits which the 1976 Law could enforce. It is noticeable that the three plants with the longest construction schedules – Krummel, Mulheim Kaerlich and Brokdorf – were among the most vulnerable to mandatory design changes because they were started in 1974 or 1975 – just in advance of the new law, and therefore liable to changes in design after site work had started. The plants started in 1976 and 1977 have been completed in the 93 to 103 month range – somewhat quicker than the average for 1974/75 plants – and this probably reflects the vendors and utilities coming to terms with the new regulatory environment created by the 1976 legal changes. This interpretation is reinforced by the fact that the reactors currently still under construction seem likely to be completed in around seven years.

However, the time trend and the Atomic Law associated with it only provides a partial explanation for the variances in construction times. (The data set is too small for useful analysis by more formal statistical means.) One explanation that is not valid in the German case – in contrast to American experience – is that of the voluntary 'stretchout' of schedules by utilities with limited financial resources or low load growth. German utilities have generally neither had the large surplus capacity nor acute cash

Table 6.11. *Construction times*[a] *of completed German commercial reactors*

Plant	Reactor type	Year of construction start	Construction time (months)	Average construction time (months)
Stade	PWR	1967	53 ⎫	
Wurgassen	BWR	1968	94 ⎪	
Biblis A	PWR	1970	62 ⎪	
Brunsbuttel	BWR	1970	83 ⎪	
Philippsburg 1	BWR	1970	112 ⎬	78
Neckarwestheim	PWR	1971	71 ⎪	
Isar 1	BWR	1972	85 ⎪	
Unterweser	PWR	1972	86 ⎪	
Biblis B	PWR	1972	59 ⎭	
Krummel	BWR	1974	122 ⎫	
Grafenrheinfeld	PWR	1975	89 ⎪	
Mulheim Kaerlich	PWR	1975	141 ⎪	
Brokdorf	PWR	1975	138 ⎬	111
Gundremmingen B	BWR	1976	96 ⎪	
Gundremmingen C	BWR	1976	102 ⎪	
Grohnde	PWR	1976	103 ⎪	
Philippsburg 2	PWR	1977	93 ⎭	

[a] Defined as period from construction start to commercial operation (handover to utility).
Source: IAEA (1978) *Power reactors in member states*, 1978 Edition, Vienna; IAEA (1985) *Nuclear power reactors in the world*, April 1985 Edition, Vienna; and IAEA (1986) *Operating experience with nuclear power stations in member states in 1984*, Vienna.

problems that have afflicted many US utilities: the predominant explanation for longer schedules over time is the interaction of technical and regulatory factors.

The most important other factor in the explanation of construction schedules, cross-cutting the time trend, is reactor type. Excluding the exceptional case of Mulheim Kaerlich (the only PWR built by BBR), KWU's PWRs have on average taken 84 months to build, compared to 99 months for BWRs. This trend is even more marked among the nine earlier reactors of Table 6.11. The five PWRs took an average of only 66 months to build, while the four BWRs took 94 months. Furthermore, Unterweser was physically constructed in only 53 months. Court action meant that it waited 21 months to go critical, followed by a further 12 months to commercial operation.

Much of the delay in the case of BWRs seems to have resulted from the design and quality problems which were uncovered during the commissioning of the Wurgassen and Brunsbuttel reactors in the mid-1970s (these problems are described in detail in the section on operating performance). In some respects the lead-times shown understate the severity of the problem since the Isar and Philippsburg reactors were shut down soon after commissioning for about a year for modifications to the pipe-work (also discussed in detail in the operating performance section).

More generally, the poorer BWR performance reflects the decision made by AEG to attempt a radical redesign of the BWR which, with hindsight, appears to have been a poor one, compounded by inadequate quality control during construction. After the Atomic Law, BWRs were also more liable to design changes, and their lower level of standardisation compared to PWR designs was also a hindrance. (See Lester, 1982, pp. 123–47.) As Siemens have asserted their control over KWU and its design of BWRs more strongly, recent experience of BWR construction is much closer to that of the PWRs.

Finally on construction times, examination of reactor size and of utility (and regional location) shows no observable effects, apart from a mild (perverse) effect for reactor size in the earlier period (smaller reactors taking slightly longer to build). The conclusion, for construction time experience, is that the regulatory changes of the late 1970s provide the most important single explanation for the differences in schedules: recent plants have been built more quickly but not in the very short periods (five years or so) that were possible in the earliest days. Reactor type has also been important: the technically less adventurous (and operationally more consistent) PWRs have had significantly shorter schedules than the more innovative and technically troubled BWRs.

On capital costs, the data are fragmentary and far from satisfactory. Of the various utilities involved in nuclear power, only RWE has attempted to provide historic data on capital costs, and the available material is highly aggregative and covers only a very limited number of plants. As in all other countries, there are substantial data on estimated future costs: however, in FR Germany, as elsewhere, estimates have been a poor guide to actual experience and these estimates are thus of limited value.

A few broad trends are, however, fairly clear. Between 1969 and 1982, pre-construction estimates of 'base' capital cost* rose at a real (inflation-adjusted) annual rate of 9% – virtually a tripling in 13 years (Marquis, 1982). This is similar to the US record but its significance is just as limited as in the US case. The reason for this is that early orders were 'turnkeys' so that the prices estimated (and paid) were probably a substantial under-estimate of actual costs incurred. Later contracts were not turnkeys: hence the extent of the increases is exaggerated, quite apart from the fact that pre-construction estimates bear no consistent (or even clearly known) relationship to actual costs. Nevertheless later pre-construction estimates

---

* 'Base' capital cost excludes interest during construction: it focusses only on the goods and services which are used to build a reactor, and ignores financing.

undoubtedly reflected substantial real increases in resources needed compared to earlier reactors.

The evidence of fairly sharp increases in inflation-adjusted capital costs is confirmed in the limited evidence on the actual costs of three RWE reactors. Evidence from the IAEA (Bennett *et al.*, 1982) via RWE suggests that Biblis A and B (1970–75 and 1972–77 construction schedules respectively) had total investment costs – including interest during construction – of approximately 500 DM/kW and 620 DM/kW. By contrast, a later, unidentified reactor of 1300 MW (thought to be Gundremmingen B or C) had a total investment cost of approximately 1620 DM/kW. Between Biblis A and Gundremmingen, a distance of 9 or 10 years in completion dates, real costs seem to have risen by over 200%. Again, qualifications are needed: Biblis was a predominantly turnkey station, so that its declared costs are probably an underestimate of true costs; second, Gundremmingen was a BWR and not yet complete at the time of the estimates quoted; and third, total investment costs reflect changes in interest rates, which may distort comparisons. On this latter point, interest rate changes were in fact limited during the relevant period in FR Germany, and although the magnitude of the total change in costs is almost certainly rather exaggerated, there is no doubt at all that German reactors have experienced very substantial real capital cost increases, mostly traceable to regulatory-induced changes and lengthening schedules.

So significant were the increases that, despite large increases in the price of German coal in the 1970s, nuclear power was beginning to lose its competitive edge over coal-firing, despite the fact that the economics of coal-fired power in FR Germany are, by international standards, quite poor. By 1982, according to Hansen (1984), the margin of nuclear generating cost advantage (i.e. including fuel, operation and maintenance etc.) over coal-firing had fallen from 87% in 1976 to a prospect of only 31% in the late 1980s. Indeed at world spot coal prices prevailing in 1983 ($45/tonne) coal-firing would have had a noticeable advantage over nuclear power: at 1986 levels of $30/tonne, nuclear power would not be uncompetitive in FR Germany. It is tempting to compare this situation with that in France: however subsidised French nuclear capital costs may have been, it is difficult to believe that fully-costed French nuclear power would not show up as rather more competitive relative to coal-firing at world coal prices than German nuclear power. Hence it is clear that the generally good technical record of German nuclear power has been purchased at substantial economic cost: in a less protected market environment than one dominated by high-priced German coal, nuclear power would have had a rather more doubtful economic rationale than has so far seemed to be the case.

OVERALL ASSESSMENT

In many respects, despite a number of errors in development, and partly through circumstance as much as design, the German nuclear industry is better equipped to survive the next decade or two than most, if not all, of its competitors.

The vendor, KWU, especially since Siemens took full control of it, has demonstrated considerable technical ability. This is illustrated by the improvements it has made to the PWR which have not been radical but resulted from the use of better materials (e.g. the steam generator), and manufacturing methods (e.g. the pressure vessel). The contrast with AEG, whose innovations were more radical and which ultimately proved to be misconceived, is marked. The emphasis on quality control has also been important. Component testing is extremely thorough and development work on smaller components such as pumps and valves has been rigorous, leading to good reliability.

KWU's decision, unlike its American competitors, not to manufacture the major components such as steam generators and pressure vessels, seems to have been justified on two counts. First, they have been able to utilise companies with pre-existing skills in the manufacture of such components, and second, it has left them with a structure better suited to the variable ordering patterns that seem an inevitable part of the nuclear business.

Also, in contrast to its American competitors, KWU carries out its own architect-engineering. This may have brought benefits in terms of shorter lines of communication; KWU may perhaps be more highly motivated to make a success of the overall project than a third-party company whose reputation is less at stake, because it is able to pass on the blame.

On the utility side, the structure is perhaps less fragmented than it may appear. Most reactor orders have involved partnerships of utilities pooling risk and experience. On a number of occasions, such as the first orders for plant and the moves towards the convoy, the utilities, particularly RWE, have taken important initiatives. In addition, utilities are often part of much larger engineering companies giving them much greater strength. For example, three of the major nuclear utilities, Preussenelektra, NWK and Isar-Amperwerke are all part of the large VEBA group.

The general absence of operator errors (with one important exception at Brunsbuttel – see earlier) and the efficient maintenance schedules – re-fuelling and maintenance is often completed in less than half of the time typically required elsewhere – reflect considerable credit on the utilities. Also, the refusal of Preussenelektra to accept hand-over of Wurgassen unit until most of the technical problems with this unit were resolved, suggests that the utilities have strong management capable of independent decision-making. However, this technological success has not been achieved cheaply.

Construction times and costs have been high, partly due to the success of objectors in delaying projects. This success seems to have been due to the requirements of the Atomic Law that plant should embody state-of-the-art technology. It is beyond the scope of this book to judge whether this has been a force maintaining the safety of plant at acceptable levels or merely an excuse to delay plant encumbering them with unnecessary safety devices. It is also difficult to say whether a more dirigiste system such as that operated by France would have been more successful in containing opposition or whether an absence of legitimate channels for protest would have made the situation even more explosive.

The safety authorities have proved tough, and in some respects, such as protection against plane crashes and gas explosions, place conditions that no other country, including the USA, requires. This is in part due to the greater population density of Germany which, necessarily, means that reactors are sited closer to population centres than most reactors in the USA. Where equipment failures have occurred, the regulatory authorities have frequently extended shutdowns beyond the time required to carry out repairs, in order to ensure that the implications of the failure were fully understood and that the repairs were thorough.

In the past, the safety regulation system has been pluralistic with several bodies such as the TUVs, individual experts and the courts having a significant impact on the requirements imposed on utilities as well as the overall federal safety regulation system. Such a degree of pluralism was unique amongst the major users of nuclear power. However, by the late 1970s it had become clear that this system was not workable and the convoy experiment can be seen as an attempt to reduce this plurality.

In some respects, although most certainly not in terms of public safety, the operator error at Brunsbuttel had elements which paralleled the TMI accident involving a misinterpretation of signals by operators who then overrode the automatic system exacerbating the original problem. This meant that the regulators were thoroughly examining the man–machine interface before the TMI accident.

The strength of opposition and stringency of regulation has had positive, albeit not always intended effects:
- delays to existing orders have meant that the nuclear component in Germany's electricity generating mix is much closer to requirements than if the forecasts of 1975 had been implemented
- the vast number of objections on safety grounds has meant that the safety of the design has been very thoroughly investigated and a large number of potential issues covered

Nevertheless there are a number of important challenges and choices the industry has to face in the next few years. Progress with the convoy system has been encouraging, but only three units have gone ahead and no second convoy is yet being assembled. Of key importance may be RWE's ability to overcome local opposition in North-Rhine Westphalia. No commercial nuclear plant has yet gone into service in this *Land* and, over the past 10 years, RWE has been continually frustrated in its plans to site units there such as those planned for Neupotz, Vahnum and Pfaffenhoffen. Unless this impasse is broken, RWE may be forced to import electricity from France if its large base of industrial customers is not to be lost to areas offering cheaper electricity. Other utilities such as NWK and HEW are now reaching the point where it would be difficult to accommodate further nuclear orders into their base load. Thus, unless *Länder* such as North-Rhine Westphalia accept the convoy principle, the scope for future orders will be limited and the convoy will have little impact on costs and construction times.

Plans for fuel cycle developments, particularly for back-end facilities, are taking shape but remain the subject of forceful opposition. Unless this opposition is overcome the lack of back-end facilities may ultimately become the most important constraint on expansion.

On technology choice, FR Germany is in a stronger position. The PWR seems likely to continue to dominate despite the recent improvement in BWRs. This is mainly on the practical grounds that if convoys are to become the typical method of ordering, it is unlikely that there will be sufficient orders to sustain convoys of both PWRs and BWRs. However, if the LWR does fall from favour, the development work on the high-temperature reactor may prove valuable.

CHAPTER 7

Canada

Introduction

The civil nuclear power programme of Canada has remained unusually single-minded in direction and objectives. Since the end of the Second World War, technology development has not been influenced by military considerations. Only one class of reactor, the heavy water moderated type reactor, has been developed with most work concentrated on the Canadian Deuterium Uranium (CANDU) type, and the main institutions – utilities, reactor vendor and regulatory authority – remain tightly knit with close personal contacts. Canada has developed the CANDU concept with little or no international back-up and its considerable success runs counter to the argument that developing complex demanding technologies can only be successfully carried out by the strongest nations or by international collaborations. In a country with few independent achievements in high technology, the CANDU reactor has become a 'flagship' of Canadian technological capability. This factor, together with the lack of any military 'contamination', has helped win the Canadian nuclear programme a measure of public approval (particularly in Ontario) that is unusual. Public opposition to nuclear power has been more limited than in other countries.

In this chapter we examine the basis of the technological achievement of CANDU reactors, and in particular how specific to a particular time and place it may have been.

THE ENERGY AND ECONOMIC BACKGROUND

The Canadian economy has experienced relatively high rates of economic growth compared with the other countries in this study: real GDP rose by an average of 5.6% per annum between 1960 and 1970, and while the recession of the mid-1970s and 1980 to 1982 had an effect, economic growth still averaged over 3.5% per annum between 1970 and 1984 (International Energy Agency, 1985a). However, in the Canadian economy, manufacturing accounts for only 15% of GDP and many of the most important economic linkages are south to the USA, and involve a relationship of dependence on the Canadian side. For a variety of reasons – including economic structure, climate and historically low energy prices – Canada has one of the world's highest per capita levels of energy use. Further, primary energy consumption continued to rise steadily through the 1970s (Table 7.1). However, a combination of increases in domestic production of coal, and nuclear and hydro power, was sufficient to outweigh the post-1973 fall in oil production. Primary energy production of all forms of energy, now including oil, is increasing steadily producing a net export of all energy forms including electricity.

In terms of final energy consumption (Table 7.2), Canada's experience in the 1973–84 period has been somewhat different to that of most OECD countries. Industrial energy use has continued to increase, and residential and transport use have shown no consistent trend. However, in common with the other countries in this study, the market shares of electricity and gas have increased (especially in the residential market) and that of oil has diminished considerably.

In energy policy, as indeed more widely, it is not always useful to think in national (i.e. federal) terms. The provincial governments have jurisdiction over all fossil fuel developments, and though federal regulation is involved in a number of areas (e.g. movement of energy resources across provincial boundaries), most energy policy decisions in practice have their origins in the individual provinces. Generally, the western provinces are net energy exporters, while the eastern provinces, especially Ontario and Quebec, are net importers, though the latter possess large hydro resources. This basic west/east split helps explain why the development of nuclear power has taken place in the east with the western provinces only showing limited interest. Canada is also rich in uranium deposits, and currently produces some 20% of the world's uranium supply.

In 1984, nuclear power accounted for about 12% of energy used in electricity generation and had already substituted for most oil and gas used for this purpose. By 1990, the proportion may have reached about 20% if units under construction are completed on schedule and overall performance is comparable with historic levels. Hydro power and cheap coal will continue to supply the bulk of electricity's primary energy requirements in Canada. It seems unlikely that nuclear power will be able to substitute to any significant extent for these sources outside the industrial eastern parts of the country.

Table 7.1. *Production and consumption of primary energy in Canada – 1973–84 (million tonnes of oil equivalent)*

	Production			Consumption			Consumption for electricity generation		
	1973	1978	1984	1973	1978	1984	1973	1978	1984
Solid fuels	11.5	22.0	39.5	15.0	23.1	34.1	8.2	12.2	21.5
Liquid fuels	94.8	76.1	83.8	84.5	89.6	69.6	2.5	3.3	1.4
Natural gas	63.2	64.3	66.0	38.7	43.5	47.9	4.3	2.2	1.5
Nuclear power	3.4	7.0	11.7	3.4	7.0	11.7	3.4	7.0	11.7
Other primary electricity	43.5	52.6	63.9	43.5	52.6	63.9	43.5	52.6	63.9
Total	216.4	222.0	264.9	185.1	215.8	227.2	61.9	77.3	100.0

Source: Energy Balances of OECD Countries, 1970–1982 and 1983–1984, International Energy Agency, Paris, 1984 and 1986.

Table 7.2. *Delivered energy consumption in Canada by sector – 1973–84 (million tonnes of oil equivalent)*

	Industry			Transportation			Residential			Other			Total		
	1973	1978	1984	1973	1978	1984	1973	1978	1984	1973	1978	1984	1973	1978	1984
Solid fuels	6.1	10.4	12.1	0.1	0	0	0.4	0.2	0.1	0	0	0.1	6.6	10.6	12.2
Liquid fuels	16.1	14.6	10.1	34.9	40.9	37.4	12.7	11.4	5.4	8.3	7.6	6.7	72.0	74.5	59.6
Gas	11.2	14.0	17.0	0	0	0.2	6.3	9.0	10.3	5.9	7.9	9.2	23.4	30.9	36.5
Electricity	9.1	10.7	13.2	0	0.2	0.2	4.5	6.8	8.6	5.1	6.5	7.9	18.7	24.2	29.9
Total	42.5	49.7	52.4	35.0	41.1	37.6	23.9	27.4	24.4	19.3	22.0	23.8	120.7	140.2	138.2

Source: as Table 7.1.

Table 7.3. *The Canadian nuclear programme*[a]

Plant name	Operator[b]	MWe (net)	Start of construction	Year of commissioning	Turbine[c] supplier
NPD 2	OH/AECL	22	1958	1962	AEI
Douglas Point[d]	OH	206	1961	1968	AEI
Gentilly 1[e]	HQ	250	1966	1972	BBC
Pickering 1–4	OH	4 × 515	1966–68	1971–73	H-P
Bruce 1–4	OH	4 × 740	1970–74	1977–79	H-P
Point Lepreau	NBP	633	1975	1983	H-P
Gentilly 2	HQ	640	1974	1983	CGE
Pickering 5–8	OH	4 × 516	1974	1983–86	H-P
Bruce 5–8	OH	4 × 750	1978–79	1984–87	CGE
Darlington 1–4	OH	4 × 881	1981–85	1988 onwards	BBC

[a] All reactors are of the 'CANDU' type (heavy water cooled and moderated pressure tube reactors) with the exception of Gentilly 1 which is a boiling light water cooled, heavy water moderated pressure tube reactor. All units were supplied by Atomic Energy of Canada Limited except NPD 2 which was supplied by Canadian General Electric.
[b] *Operators*   OH – Ontario Hydro   AECL – Atomic Energy Canada Limited
            HQ – Hydro-Quebec   NBP – New Brunswick Electric Power Commission
[c] *Turbine suppliers*   AEI – Associated Electric Industries (UK)
            BBC – Brown Boveri (Switzerland and FR Germany)
            CGE – Canadian General Electric (Canada)
            H-P – Howden-Parsons (Canada/UK)
[d] Douglas Point last produced power in May 1984 and was permanently closed in summer 1984.
[e] Gentilly 1 last produced power in March 1979 and was permanently closed in autumn 1981.
Source: SPRU nuclear power plant data base.

THE DEVELOPMENT OF NUCLEAR POWER

The evolution of nuclear power in Canada can be conveniently split into four periods.

The first, pre-commercial phase, dates from 1945 until about 1966. The US McMahon Act, which denied enriched uranium to the USA's wartime Manhattan project partners, led Canada, like the UK and France, to develop technology based on a moderator which allowed the use of natural uranium. Unlike the European countries, which mostly chose graphite as the moderator, Canada opted for heavy water moderated reactors as its main focus. Initial development was slow and the first prototype NPD 2 (22 MWe) (Table 7.3) was not completed until 1962, to be followed by the much larger Douglas Point (208 MWe) in 1966. A similar design was incorporated in the two-unit sale to India (RAPP 1 and 2) (Table 7.4).

The second phase from 1966 to 1974 marked the beginnings of commercialisation. In 1966, construction started on both a demonstration light

Table 7.4. *Exports of Canadian reactors*

Country	Plant name	MWe (net)	Start of construction	Year of commissioning	NSSS[a] supplier	Turbine[b] supplier
India	RAPP 1&2	2 × 207	1964/68	1973/81	AECL/Dept AE	EE
Pakistan	Kanupp	125	1966	1972	CGE	Hitachi
Argentina	Embalse (Cordoba)	600	1973	1983	AECL	Ansaldo
South Korea	Wolsung	629	1977	1983	AECL	Parsons
Romania	Cernavoda 1–4	4 × 629	1980–85	1989–92	AECL	GE

[a] *NSSS suppliers*
    AECL – Atomic Energy of Canada Limited
    Dept A.E. – Department of Atomic Energy (India)
    CGE – Canadian General Electric
[b] *Turbine suppliers*
    EE – English Electric (UK)
    Hitachi – Hitachi (Japan)
    Ansaldo – Ansaldo (Italy)
    Parsons – C. A. Parsons (UK)
    GE – General Electric (USA)

Source: SPRU nuclear power plant data base.

water cooled reactor (Gentilly 1) and, more important, the heavy water cooled Pickering units, a four-reactor station for Ontario Hydro in direct developmental line from NPD and Douglas Point. A scaled-up CANDU design, of some 50% greater capacity than Pickering, was ordered, also by Ontario Hydro, in the 1970–74 period (the four Bruce units), and export orders were won in Pakistan and Argentina.

The third period, beginning in 1974, was one which promised substantial domestic expansion but serious export problems, following the Indian nuclear explosion of 1974 which used materials produced by technology imported from Canada. In the wake of the first oil crisis, high official forecasts of future nuclear needs were made, and although these were quickly seen to be exaggerated, eight more units were ordered for Ontario Hydro at the existing Bruce and Pickering sites. In addition, the first domestic commercial orders outside Ontario were placed by New Brunswick Power (Point Lepreau) and Hydro-Quebec (Gentilly 2). These were also the years in which extremely encouraging operating performance started to emerge in the four Pickering units.

The fourth period, from 1979 to the present, has been, as for most nuclear industries, a much harder one. Expectations of future need were further reduced and only four units for the Darlington site were ordered in this period. Since 1977, the only sales achieved outside Ontario have been the four units sold to Romania and the possible sale of a single unit to Turkey. Whilst Canada has competed vigorously in all the export markets that have arisen, a principal concern has been the maintenance of a continuing option to build CANDU, as and when necessary. On the operating side, the good early performance of Pickering has been exceeded by the first four units at Bruce, but the reputation of CANDUs for very high performance has been tarnished in the last two or three years by a significant technical problem at the early Pickering units which may have implications for all CANDUs.

THE KEY INSTITUTIONS

The institutional structure surrounding civil nuclear power is dominated by publicly owned and controlled organisations. Electric utilities are organised on a provincial basis and each of the three utilities which operates nuclear plant, Ontario Hydro, Hydro-Quebec and the New Brunswick Electric Power Commission is owned by its respective province. Of these, Ontario Hydro dominates and the other two utilities currently own only one unit each.

The other institutions are owned/controlled at the federal level although all are based in Ontario. These are the vendor and state nuclear research organisation, Atomic Energy of Canada Limited (AECL) and the safety regulator, the Atomic Energy Control Board (AECB). This concentration of organisations in Ontario, mainly in Toronto and Ottawa, has allowed considerable interchange of personnel between organisations.

Ontario Hydro

By world standards, Ontario Hydro is a large electric utility operating more than 20 GW of plant. Although much of this plant is hydro-electric, Ontario Hydro has a substantial history of operating thermal units, both coal-fired and oil-fired. Despite being owned by the province of Ontario, it generally operates at 'arm's length' from the government and is financially self-supporting. Whilst Hydro-Quebec and New Brunswick Power have both ordered CANDUs, they have not played any central part in nuclear development.

AECB

The AECB was set up in 1946 by the federal government and has responsibility for a wide range of activities including research, but, since 1954, its activities have been restricted mainly to regulation although recently its sponsorship of research has been increasing. The AECB reports to the federal parliament through the Minister for Energy, Mines and Resources.

The system of licensing plant is a relatively simple three stage procedure (Nuclear Energy Agency, 1980); the first step is site approval, the second, construction licence and the third, operating licence. In these proceedings, the AECB is advised by the Reactor Safety Advisory Committee (RSAC), which has 15 members drawn from engineers and scientists involved in nuclear power together with technical representatives of government agencies. The site approval step is not formally required but is current practice. A summary description of the plant is provided including its basic features and local environmental information. Recently, the AECB has begun to hold public hearings at this stage but the length and scope are at the discretion of the AECB itself and it is not clear what requirement there is for the AECB to act on the findings of the hearing, or indeed whether the hearings are in fact legal. In the future it may also be necessary to meet the requirements of the provincial environmental regulating agencies.

Once site approval has been granted, the design is further developed and a construction licence may then be applied for. This application must include a safety report. However, the design is not finalised at this stage and after the licence has been granted by the AECB and construction is in progress, meetings take place with the applicant to resolve safety issues.

When construction is near completion, the applicant requests an operating licence. A final version of the safety report is tabled along with information on the staffing and procedures that will be used in the operation of the plant. Once the plant is in operation, the AECB can impose limitations on the output of the plant or require it to be shut down if they are not satisfied with any aspect of the plant. The regulatory process as a whole gives the AECB wide discretionary powers and is almost entirely conducted in private.

AECL

Since 1968, when Canadian General Electric (CGE) withdrew from reactor sales, AECL, through its reactor design division, CANDU Operations, has been the sole supplier of CANDU technology in Canada, though Ontario Hydro has itself developed some independent capability over the years. Essentially AECL, like Kraftwerk Union in Germany, supplies software in terms of designs and, importantly, computer control systems. Hardware orders are sub-contracted to large general engineering companies, often subsidiaries of US firms. The main suppliers in the past have been Canadian Babcock & Wilcox and Canadian Foster Wheeler for steam generators, Canadian Vickers, Chase Brass and Dominion Bridge Company for reactor parts, Canadian General Electric for fuel and fuelling machines and Canatom, a subsidiary of three Canadian engineering companies, for nuclear engineering. Unlike KWU, however, AECL is not part of a large group, it does not supply turbine generators (a major incentive for KWU to maintain a nuclear capability) and the terms of its constitution do not allow it to diversify into related industrial areas. AECL has, however, been responsible for the two heavy water plants at Glace Bay and Port Hawkesbury. The combination of serious technical difficulties and the virtual collapse of heavy water demand has meant that AECL has experienced severe financial losses in its heavy water operations. Federal pressure aimed at maintaining employment delayed the closure of these plants until 1985 despite the surplus capacity in heavy water supply. AECL is also, through a separate arm, the state nuclear research organisation.

THE PRE-COMMERCIAL PERIOD (1945–66)

Alone among the partners in the wartime Manhattan project, Canada chose after the Second World War to renounce the development or use of nuclear weapons. Consequently, whilst the other three countries in this study have designed and built commercial power reactors based on military designs (submarines in the case of the USA, and plutonium production in France and the UK), Canada's reactors were designed purely for power generation purposes. The US McMahon Act, which was passed immediately after the war, limited Canada, France and the UK's access to enriched uranium and thus effectively prohibited the use of light water as a moderator, leaving heavy water and graphite as the most attractive choices. Interest in heavy water reactors has been considerable, and several countries, notably Germany, France, Switzerland, Norway, Czechoslovakia, Italy, Sweden and the UK, have built prototype heavy water moderated power reactors using a variety of coolants (carbon dioxide, boiling light water and boiling heavy water). Canada itself has built a prototype organically cooled reactor and a semi-commercial size boiling light water cooled reactor, but, of the heavy water designs developed, only the CANDU reactor and a German design which both use pressurised heavy water as both coolant and moderator have been built commercially.

The pace of development was slower than in the USA, France and the UK and the first prototype (NPD) was not completed until 1962, after about 10 years' design and construction work by AECL and CGE. AECL had by then started work on the much larger Douglas Point unit (208 MWe) and this was completed in 1966, by which time the first full-scale units at Pickering were under construction.

THE BEGINNINGS OF COMMERCIALISATION (1966–74)

In the mid-1960s it was decided to build a demonstration plant using a slightly different technology from that of the basic CANDU. Instead of using pressurised heavy water as the coolant, boiling light water was chosen. This decision was a result of worries about shortages of heavy water. It was expected that this would also reduce costs by allowing a smaller heavy water inventory (about 40% less). The use of a 'once-through' steam system would also make design and construction simpler. The station (Gentilly 1) was built and financed by AECL with the expectation that Hydro-Quebec would purchase the plant once it was operating satisfactorily. In fact the plant never did operate satisfactorily and was not purchased by Hydro-Quebec, being permanently shut down in 1981.

In export markets (see Table 7.4), some success was achieved with sales of reactors to India by AECL using the Douglas Point design, and to Pakistan by CGE. In 1968, CGE withdrew from the reactor sales market to concentrate on specialised services.

By the early 1970s, the CANDU was achieving commercial status. The reactors at Pickering had been completed and those at Bruce were under construction. Plans were advanced for further similar tranches of plant at Pickering and Bruce to double capacity at these sites. The Bruce design was to be developed for a further tranch of four units to be built at the Darlington site. All these developments were within Ontario, but inroads were also made outside Ontario with sales (single units) of a different design to Hydro-Quebec (Gentilly 2) and New Brunswick (Point Lebreau). Export sales to Argentina (Cordoba) and, later, South Korea (Wolsung) were achieved using this latter design.

FULL COMMERCIALISATION (1974–79)

The domestic prospects for the Canadian industry seemed encouraging following the first oil crisis. In terms of technology, the Pickering units were performing well and had acquired a reputation for reliability and engineering excellence. Heavy water production was beginning to increase with the opening of the Port Hawkesbury and Bruce A plants. This obviated the need to import heavy water from the US Savannah River plant which had become necessary following the failure of the first full-scale heavy water plant constructed at Glace Bay, Nova Scotia. Fears of a shortage of heavy water had been seen as a possible constraint to nuclear expansion and were instrumental in the decision in 1966 to use light water as the coolant in the Gentilly 1 plant. By 1975 it was fairly clear that this prototype had not been successful. However, the slower than expected pace of nuclear capacity growth and the success of the basic CANDU meant that there was little concern about this failure. AECL did suffer heavy financial losses on this prototype because Hydro-Quebec did not take up the option to purchase.

The government saw an apparent need for an enormous expansion in nuclear capacity by the turn of the century. Donald Macdonald, the then Federal Minister of Energy wrote (1974, p. 475):

Much of our industry is energy concentrative and we have one of the highest per capita rates of electricity consumption in the world. Indications for the future are that these trends will continue if not accelerate, as the industrialization and urbanization of the country continue.

Nuclear power is destined to play an important role in this process . . . nuclear power accounts for less than 5 per cent of the electricity produced in Canada. By the end of the century this will have grown to 50 per cent. To achieve this rate of growth, some 130,000 MW(e) of nuclear capacity will have to be added in the next twenty-five years.

Like most other contemporary forecasts of nuclear futures this proved to be a gross over-estimate.

The major difficulty in the mid-1970s concerned reactor exports, in particular the risk of weapons proliferation resulting from the spread of nuclear weapons materials and technology. CANDU reactors pose particular problems in this respect for two reasons. First, the spent fuel from

CANDU reactors tends to contain high purity ^{239}Pu (i.e. contamination with ^{242}Pu is lower than with light water reactors) so that they are a particularly good source of weapons-grade plutonium. Secondly, the continuous refuelling that is a feature of the CANDU makes safeguarding difficult since fuel elements can be more easily subverted without detection by the safeguarding authorities. Concern was triggered by the nuclear explosion detonated by the Indian Government in May 1974, using plutonium derived from the Cirus research reactor supplied by Canada. All support for the Indian programme was immediately withdrawn and in 1976 support to Pakistan was also withdrawn. At the time of the Indian explosion, Canada had begun to achieve some success in export markets, including a sale in 1971 to Argentina and negotiations with South Korea. Like other countries at that time, Canada was fairly relaxed in its requirements for safeguards, but the situation then changed radically. The South Korean deal was delayed pending agreement by South Korea to a number of conditions. From 1975, Canada refused to sell reactors to countries not applying full-scope IAEA safeguards. This ruled out a number of Third World Countries with unsafeguarded facilities. In addition, AECL was under a cloud domestically following the disclosure of the payment of bribes to Argentine officials. These factors meant that from 1975 until about 1978, the CANDU was severely handicapped in competitions for export sales.

Overall, since the main fears concerned the sufficiency of human resources to meet Canada's own projected nuclear requirements, the export limitations did not cause as much concern as a more realistic appraisal of domestic demand might have suggested.

THE RECESSION (1979 ONWARDS)

As in almost all countries, the grandiose plans of the immediate post-oil crisis period soon foundered as the expected early return to the levels of economic growth achieved in the 1950s and 1960s failed to materialise. The recession in plant orders has not hit Canadian industry as hard as US vendors. The reason for this is that the pattern of ordering was smoother in the 1970s and four units at Darlington were ordered after 1980. This meant that the number of CANDU units actively under construction worldwide by the mid-1980s was only a little less than when a dozen units were being built in the mid-1970s. The major difference between now and the mid-1970s is that much of the hardware for these orders has already been shipped and the prospects for new orders either for the home market or for export appear limited.

This decline and the accompanying difficulties of maintaining critical manufacturing resources and skills has been recognised since the late 1970s.

Electricity demand in Canada has not grown sufficiently to warrant further home orders and the construction schedules for the Darlington units have been stretched out in recent years. In fact there are now increasing doubts as to whether units 3 and 4 of this development will be built. Attention has switched to building reactors to supply electricity to the USA, particularly the East Coast, via a second unit at Point Lepreau. In fact electricity demand in New Brunswick had fallen so far short of expectations that it was possible to contract 335 MW out of the 630 MW of Point Lepreau 1 to utilities in New England. Whether it will be possible politically and financially to build a reactor dedicated to the export of electricity, even allowing for the strong support of New Brunswick Power and the New England utilities, remains to be seen.

In the export field, high hopes of achieving sales to Japan, Mexico and a second unit to South Korea have come to nothing despite assiduous marketing effort. Amidst much acrimony about the extent of proliferation safeguards required, a sale to Argentina was lost to KWU's design of heavy water reactor in spite of a much lower AECL bid. There are still hopes for Turkey and Egypt but financing reactor construction in developing countries is a very severe problem for all vendors. A novel solution to these problems has been proposed in AECL's bid for a sale to Turkey. This would involve AECL (in collaboration with Ontario Hydro) building and operating the plant and selling its output to the Turkish electric utility.

An agreement to supply reactors to Romania, with an initial order for four units, has also provoked controversy. There is concern in Canada about the extent to which Canadian knowledge and expertise will be 'given away' with this order. The fear is that the agreement will give so much knowledge away that any subsequent Romanian CANDUs will bring little benefit to Canada. This controversy has also arisen with respect to attempts to gain further sales to South Korea. In practice it is now difficult to sell any reactor without a significant component of technological transfer: any expectation of a long and continuing process of hardware supply to a recipient is essentially unrealistic.

Various aspects of the nuclear industry have been under close public scrutiny in the past five years through a number of reports and inquiries. The most important of these was probably the Ontario Royal Commission on Electric Power Planning's report on nuclear power in Ontario (1980) – the so-called Porter Commission – which was set up in 1977. The report issued in 1980 foresaw a dramatically lower requirement for nuclear plant in Ontario by 2000 than was then widely forecast although its expectation of a need for a 3400 MW station to follow Darlington now appears unrealistic. The status of the report was somewhat clouded by statements made by Professor Arthur Porter, its chairman, some six months after its publication. In evidence to the Commons Select Committee on Alternative Energy & Oil Substitution, it became clear that Porter's personal position differed from that adopted by the Royal Commission which had advocated a balanced development of energy sources including renewables. Porter advocated a strategy in which nuclear power was more central and he urged the early construction of additional CANDUs including some dedicated to the export of power to USA. He played down the role of renewables (see *Nucleonics Week*, 1980, pp. 7–8).

The whole of the Canadian nuclear industry has been under close government scrutiny both at the federal and provincial level over the past few years, with particular emphasis on finding an acceptable size and structure for the plant supply industry. The plant supply industry itself commissioned a consultant's report on this subject, the SECOR report (Lortie & Schweitzer, 1981), which made recommendations on issues such as international marketing, component manufacture and non-proliferation. Whilst most of the report is well-argued, the dearth of sales has left little scope to implement its ideas.

ECONOMIC PERFORMANCE

Operating performance

An assessment of the operating performance of the Canadian CANDU reactors must be based almost entirely on the first four reactors at each of the Pickering and Bruce sites. The other reactors are all still in the early years of operation.

The performance of the CANDU reactors is summarised in Table 7.5 and

Table 7.5. *Performance of mature CANDU reactors by time period*

	Pickering		Bruce		Point Lepreau/Gentilly 2	
	Capacity factor	No. of reactor years	Capacity factor	No. of reactor years	Capacity factor	No. of reactor years
1976–79	87.5	16	—	—	—	—
1980–82	85.9	12	89.6	9	—	—
1983–85	50.5	12	92.9	12	—	—
1986	48.8	5	82.8	4	87.9	2

Source: SPRU nuclear power plant data base.

Table 7.6. *Performance of CANDU reactors by unit age*

Age	Capacity factor (no. of reactor years)		
	Pickering	Point Lepreau/Gentilly 2	Bruce
1	72.2 (7)	75.4 (2)	77.6 (6)
2	64.1 (6)	77.4 (2)	82.4 (5)
3 onwards	73.3 (50)	87.9 (2)	90.1 (25)

Source: SPRU nuclear power plant data base.

Table 7.6 and shown in greater detail in the Appendix to this chapter. The main points to emerge from these analyses are:
– with the exception of recent performance at the Pickering sites (see below) the reactors at both sites have achieved a very high and consistent level of performance
– the settling-in period has been short and early-year performance has improved with successive generations (on early evidence this seems likely to apply to the second tranches at Pickering and Bruce)
– no significant derating has been necessary
– the reactors do not suffer significantly from operating losses, i.e. the units operate at, or very close to, full rating whenever they are available for service
– no shutdowns have occurred ascribed solely to regulatory action
– shutdown frequency is low, suggesting good reliability

Performance, regulation and problem-solving

Two technical areas deserve further elaboration in this context. These are pressure tube and steam generator performance. The use of pressure tubes is one of the unique (in terms of commercial reactors) features of the CANDU technology and in this area Canada has not been able to benefit from international experience. By contrast steam generators are incorporated in all PWRs. The way in which problems with these components have been handled is also a useful indicator of the way in which the regulatory system operates and tests the idea that the regulatory system in Canada may be unduly incestuous and biased towards the interests of the operator at the expense of accountability and possibly public safety.

The pressure tubes. The three main reasons for using a number of pressure tubes rather than a single pressure vessel were intrinsic safety, Canada's relative lack of metallurgical skills, and the potential for plant life-extension they gave. Pressure tubes were thought to be safer because in the case of a tube failure, the tube was expected to leak before it ruptured and could be withdrawn and replaced with minimal disruption. A single pressure vessel, if it contained a crack above a certain critical size, would rupture without warning. The manufacture of large pressure vessels requires considerable skills and manufacturing capability which are available in few countries whereas pressure tube manufacture is less demanding. In fact the use of heavy water as moderator and natural uranium would require a larger pressure vessel to achieve the same electrical output as a comparable PWR and would probably limit the size of unit to about 700 MW. The potential for life-extension arises because it was expected to be relatively easy to replace pressure tubes whereas the replacement of a pressure vessel was thought to be infeasible. After more than a decade of experience it is useful to review these expected advantages and to investigate whether the lack of international back-up has been a significant disadvantage.

In fact the pressure tubes have been the most problematic area of the design. The three most significant problems to have arisen are pressure tube leaks at Pickering 3 & 4 discovered in 1974, a continuing problem of unanticipated pressure tube creep and the pressure tube rupture in Pickering 2 in 1983.

The pressure tubes were designed to have a 20-year life with a firm expectation that they could then be replaced to give the unit a further 20 years' life. In the first two commercial units (Pickering 1 & 2), the tubes were made of Zircaloy (almost pure zirconium) whilst subsequent units have used a zirconium/niobium alloy for reasons of improved neutron economy, and extra toughness.

The heavy water leaks detected at Pickering 3 and subsequently found in unit 4, were the first serious problem at a commercial CANDU. The leaks were caused by an error in manufacturing and resulted in lengthy shutdowns (of eight months at unit 3 followed by an 11-month shutdown at unit 4) for testing of the tubes. In all, of the 390 tubes at each unit, 69 were replaced in total. No other units were affected.

The second problem to be detected was pressure tube 'creep' (elongation), a phenomenon induced by heat and radiation. The 6 metre long tubes extend by about 4–5 mm per year, somewhat more than originally expected. The design tolerance built into the early units to accommodate this creep was found to be only sufficient for about 15 years' service, at which time the tube would begin to distort and suffer undue stress, and would need replacement. Allowance for 25–30 years' creep was built into Bruce 4 and all subsequent units so that the affected units were Pickering 1–4 and Bruce 1–3.

Initially it was feared that the pressure tubes would have to be replaced at 15 years – a very expensive job expected to take 15 months or more – but subsequently it was decided that a repositioning operation was feasible and this would extend the pressure tube life to the design length of 25–30 years. This repositioning, known as 'refab', would take a total of four months, split into two phases to avoid shutdowns during periods of high electricity demand.

The pressure tube rupture at Pickering 2 (in 1983) caused the 'refab' programme to be re-evaluated. It was the most serious incident at a CANDU unit yet and, for some time, called into question the generic safety of CANDUs. The incident occurred at full power after 342 days of continuous operation when a pressure tube ruptured without a detected prior leak. It had previously been expected that the pressure tubes would leak before they ruptured. The plant's systems all functioned correctly and the plant was shut down according to design expectations and without the need to use the emergency core cooling system, and without significant release of radiation either to the environment or within the containment.

A number of factors contributed to the accident. Garter springs, which hold the pressure tubes in place and separate them from the calandria – the framework that holds the pressure tubes – had become displaced over a period due to a design error, resulting in contact between the pressure tube and calandria. This had led to local cooling of the pressure tube and the formation of blisters of zirconium hydride over a period of nearly 10 years. This in turn resulted in embrittlement and cracking, and finally rupture. Zirconium/niobium is not prone to the embrittlement and so only the first two units Pickering 1 & 2 were susceptible to this problem. After inspection of tubes at these two units, it was found that blistering due to displaced garter springs had affected a significant number of tubes. Ontario Hydro had initially hoped to continue to operate these units until replacement tubes were available and the shutdowns could be conveniently phased. They argued that the smoothness of the shutdowns following the rupture indicated that continued operation was safe. The AECB was unwilling to allow further operation and it was decided in March 1984 that the two Pickering units would be completely retubed using zirconium/niobium. This operation was completed in 1987.

The steam generators. This is an important area because of the chronic problems suffered by the steam generators in other reactors, particularly those of Westinghouse design. The materials, design and manufacturers have not remained constant although all the steam generators for Canadian installed units were built by Babcock & Wilcox. For each of the Pickering A units, 12 boilers were used containing about 31 000 Monel tubes, Pickering

B units will be essentially identical. For Bruce A units, eight boilers were used with a total of approximately 34 000 Inconel–600 tubes – the material used in Westinghouse units – and this design will also be repeated for the Bruce B unit. The CANDU 600 units such as Gentilly 2 and Point Lepreau have four boilers each containing about 1400 Incoloy-800 tubes – the material used in Kraftwerk Union PWRs.

The one major problem so far encountered, occurred during the construction phase of the Pickering B units and the units of about 600 MW (the CANDU 600 units). These units used Babcock & Wilcox steam generators – the South Korean Wolsung unit's steam generators were supplied by Foster Wheeler and were not affected. The problem arose as a result of an error in heat treatment during manufacture which led to indentations and deformation of the tubing and the supports. A complex and lengthy repair job had to be carried out which delayed the commissioning of the affected units by up to two years. After lengthy negotiations, Babcock & Wilcox agreed to pay for most of the repair costs, but the cost of delay – an order of magnitude greater at about C$500m – has been borne by Ontario Hydro.

Further examples of the way in which safety matters are resolved between the AECB, AECL and Ontario Hydro relate to the prototype reactors at Douglas Point and Gentilly. In 1977, the AECB required that the output of Douglas Point be reduced because of doubts about the emergency core cooling system (ECCS). Modifications to the ECCS were carried out in an eight-month shutdown and completed in October 1980. AECL and Ontario Hydro then applied for a full power licence but this was not granted until November 1981 after further analysis of system performance had been supplied.

Although the unit operated well in 1982, it was decided in 1984 that it should be closed down as the cost of repairs that would soon be required, particularly pressure tube replacement, was not justified by the value of its output, especially with the second tranche of Pickering and Bruce units beginning to enter service.

Gentilly 1's operating record was very poor and in its ten years' life, it only produced power on 181 days, most recently in 1979. Hydro-Quebec (the operator) and AECL (the owner) were unable to produce plans to convince the AECB that the plant could be brought up to an acceptable safety standard. In 1981 its operating licence was revoked and the unit will not be operated again.

Overall, whilst it is clear that Ontario Hydro and AECL do enjoy a close relationship with the AECB, the AECB has, on a number of occasions, demonstrated that it is not captive to the industries' wishes. What the impact of a more restrictive, less co-operative style of regulation would have been on the operation of CANDUs is impossible to assess. A more public and accountable regulatory system might have allowed this to be tested more effectively.

Capital costs and lead-times

As argued earlier in Chapter 2, definitive assessment of performance, even within one country, is considerably more difficult in the case of lead-times and (especially) capital costs than it is for operating performance. For lead-times the problems are less acute because, if attention is confined to *construction* lead-times, there is a reasonably unambiguous definition of the parameter to be measured, and a standard unit (time) in which to measure it. The main difficulty is that in recent years, in Canada as elsewhere, electricity demand and financing problems have meant that utilities have not always sought to minimise construction times.

Construction times. Table 7.7 shows construction times for all reactors either commissioned or close to completion. The Pickering 1–4 units were completed close to schedule and experienced few construction problems. Despite their larger size the Bruce 1–4 units did not on average take any longer to construct than the first four Pickering units: indeed, but for pressure-tube cracking problems on the first two units, they might have been built more quickly. (Much of the detail in this and the next section is drawn from Yu & Bate, 1982.)

There was a marked change in experience on reactors on which construction started in the mid-1970s. All six took at least a year longer to build than any previous Canadian reactor, and on average they took some three years longer to build than Pickering 1–4 or Bruce 1–4. Gentilly 2 and Point Lepreau might have been expected to take longer to build than the more or less contemporary Pickering units 5–8, because of the relative inexperience in nuclear construction on the part of the utilities and local labour forces involved. In practice, however, the construction times at Pickering were rather worse than at Gentilly 2 and Point Lepreau. Precise explanations for the length of construction times at Gentilly and Point Lepreau are not available. However, steam generator problems (similar to these experienced at Pickering 5–8) caused some minor delays, as did new regulatory requirements such as the need to install a high-pressure emergency core cooling system.

The long construction schedules at Pickering 5–8 were more unexpected: Pickering stage two was basically a repeat of the earlier Pickering station, built by an experienced utility and labour force. However, as a footnote to Table 7.7 indicates, units 6–8 may have been built somewhat more quickly than shown. Of the three years longer taken at Pickering 5–8 compared with

Table 7.7. *Construction times of Canadian reactors*

Reactor	Year of construction start	Construction time[a] (months)	Average construction times
Pickering 1	1966	65	
2	1966	70	68
3	1967	63	
4	1968	75	
Bruce 1	1970	78	
2	1970	80	69
3	1973	60	
4	1974	59	
Gentilly 2	1974	114	
Point Lepreau	1975	92	
Pickering 5	1974	94[b]	113
6	1974	115[b]	
7	1974	126[b]	
8	1974	139[b]	
Bruce 5	1978	80	
6	1978	77	81
7	1979	81	
8	1979	88[c]	

[a]Construction time refers to the period between construction start (generally, first structural concrete) and commissioning (handing over to owning utility).
[b]At a four-reactor site, it would be unusual for construction to start on all reactors in the same year, as the IAEA data claim for Pickering 5–8. This means that the construction times for Pickering 6, 7 and 8 may well have been shorter than the IAEA data suggest.
[c]Bruce 4 has not yet been commissioned. The time given is therefore an estimate, with a potential bias towards being too low.
Source: Lester, R. K. (1982) Nuclear power plant lead times in I. Smart (ed), *World nuclear energy: toward a bargain of confidence,* Johns Hopkins, Baltimore; International Atomic Energy Agency (1985) *Nuclear power reactors in the world,* Vienna, April 1985.

1–4, a full year is accounted for by a deliberate 'stretchout' in schedule, following capital constraints imposed by the Ontario Government in 1976 and reductions in load forecasts. Most of the remaining two years was filled by a delay due to the faulty manufacture of the original Babcock & Wilcox steam generators (see section on operating performance).

The second stage of Bruce (like Pickering 5–8 essentially a repeat of the first four units on the site) had schedules roughly a year longer than Bruce 1–4. This extra time is accounted for mainly by some regulatory interventions and changes in project scope, together with a stretchout because of slow growth in electricity demand.

Overall, this lead-time experience shows little evidence for any long-term trends in construction times over the twenty years of Canadian reactor construction. There was certainly a serious deterioration in construction performance among reactors on which construction started in the mid-1970s, but subsequent experience at Bruce suggests that this was related to a specific technical problem (steam generator manufacture), to the entry of new and inexperienced utilities, and to a slowing in load-growth expectations. The Darlington units may well have long construction schedules but this seems likely to be due overwhelmingly to sharp falls in expectations of load-growth resulting in voluntary stretchouts, rather than to endemic technical or economic problems in the industry. The Canadian construction experience is therefore in marked contrast to the substantial overruns in construction schedules in the USA.

Capital costs. Until the early 1980s, there was a well-established conventional wisdom – prevalent in Canada as elsewhere* – that CANDU capital costs would inevitably be higher than those of the rival light water reactor systems. Two reasons were given: first, that CANDUs had lower power densities and would therefore have higher capital costs due to their larger physical size and therefore material requirements: and second, that the heavy water inventory, equivalent to some 20% of the 'dry' capital cost, would push up initial CANDU costs yet further. More recently, this wisdom has been challenged, and whatever the difficulties in establishing exact figures, it can now plausibly be argued that in given regulatory conditions, CANDUs may not carry any significant capital cost penalty compared to LWRs. (See, for instance, Lortie & Schweitzer, 1981, and Kim & Chung, 1979.)

Whilst international and inter-reactor type comparisons are not the main focus of this capital cost analysis, this major change clearly suggests a significantly different Canadian experience compared with the USA in capital costs as well as in construction times.

Capital cost data for Canada are available only from Ontario Hydro. They appear to be prepared in a way that is consistent, and consequently provide a basis for establishing the movement in capital costs through time at the aggregative station-wide level.†

---

* For instance, CANATOM studies done in 1977/78 showed the 'wet' CANDU capital cost (including heavy water inventory) as some 40% more expensive than a PWR. Like many other studies, this was an engineering-based estimate of prospective future costs.

† Because many of the expenditures for a multi-unit site are common, apparent differences in cost between units on such a site might not be meaningful.

Table 7.8. *Capital costs for Ontario Hydro CANDU stations (1982 C$/kW)*

	(1) Design and construction cost only		(2) Inclusive cost (with heavy water, commissioning and monetary IDC)	
	$/kW	Index	$/kW	Index
Pickering 1–4	852	100	1194	100
Bruce 1–4	878	103	1341	112
Pickering 5–8	1145	134	1856	156
Bruce 5–8	1019	120	1533	128

Source: Column (1) from personal communication, W G Morison, Director of Design and Development for Ontario Hydro, 7 September, 1983.
Column (2) from Leonard L. Bennett, Karousakis & Moynet (1982) *Review of nuclear power costs around the world*, paper presented to the IAEA Conference on Nuclear Power Experience, Vienna, 13–17 September, 1982. (IAEA-CB-42/76.)

As is argued in Chapter 3, movement in real historical costs through time is in any case the most satisfactory form of national capital cost analysis; comparison of results with estimates begs questions about the meaning of estimates, while forward projections are of little value given the historical record of mis-estimation.

Table 7.8 shows real inflation-adjusted capital costs for the four Ontario Hydro stations that are either complete or nearly so. The first column of figures shows the results in terms of design and construction costs alone (excluding heavy water inventories and commissioning costs): the second column includes heavy water and commissioning costs, plus the impact of interest during construction (IDC). Whilst this latter set of figures is more comprehensive, it suffers from the disadvantage that interest rates fluctuated substantially during the construction period covered, and so the differences in capital cost shown are partly a reflection of the changing *price* of capital. Ideally, interest during construction should be included, but at a standard discount rate to overcome this interest rate fluctuation problem. Unfortunately the data do not allow such a manipulation, and so the two columns of figures provide incomplete, though certainly valuable, guides to real capital cost movements.

Table 7.8 shows that unit capital costs have clearly been higher in the repeat stations, Bruce 5–8 and Pickering 5–8, than in the originals. While this is a familiar result in nuclear power generally, it is nevertheless important to stress, because in most technologies repetition should, via learning, reduce real capital costs. Given that Canadian costs have escalated less rapidly than in other countries studied in this book, learning may well have taken place, but has been outweighed by other cost-increasing factors. The most general factor appears to be a continuous tightening in regulatory standards, leading to increased project scope over time. Instances of this tendency include the high-pressure emergency core cooling systems already referred to, enhanced seismic qualifications and extensions to the fuel handling system. However, it would be wrong to conclude that regulation is the only force involved. Apart from the particular steam generator problems at Pickering 5–8, there have also been autonomous design changes leading to higher capital costs, including re-design of the steam drum and major changes in station layout.

The exceptional experience at Pickering 5–8 means that no consistent upward trend over time emerges from the data in Table 7.8. Comparisons can be made between the two Pickering stations and between the two Bruce stations. Over the nine-year period beween construction start dates, at Pickering, real costs have increased by some 3% per annum (considering design and construction costs alone) or some 5% (including heavy water, commissioning and our imperfect IDC measure). By contrast, the Bruce costs rose over the equivalent seven-year gap by only around 2% per annum on either basis of measurement. Given that the second Pickering station may not be representative, and that the figures including IDC may well be inflated in later years because of rising interest rates, the general conclusion appears to be that real unit capital costs have been increasing in Canada at a rate little higher than an average of 2% per annum.

Opposition to nuclear power

It has been possible to discuss the Canadian experience of nuclear power without reference to public opposition, because opposition has neither been widespread nor very effective in the narrow sense of disrupting or halting the programme.* Of course, as in other countries, opposition has meant that decisions on safety and rate of ordering have been subjected to far greater scrutiny than they might otherwise have been. Nevertheless the limited scope of opposition is, on the face of it, surprising, given the close and non-confrontational relationship between the regulators and the industry, the lack of any legal or public access for intervenors, and the close proximity to the much more turbulent US experience. These three factors might have been expected to provoke significant political hostility to nuclear power.

---

* There is a long history of public disquiet over the environmental and safety impacts of uranium mining. However, the setting up of public inquiries such as that into the proposal to open a uranium mine and mill at Cluff Lake appear to have been successful in meeting this concern.

There would seem to be a number of reasons for the limited scope of opposition:
- nuclear power has been developed on very few sites, most of them remote. The history of opposition in the USA has been that local opposition to specific sites has been a powerful force in anti-nuclear organisation
- it has been possible to present the CANDU as a reactor system completely separate from the American LWR, and this has helped isolate it from the reactor safety issues publicised in the USA. With the exception of the Pickering pressure tube rupture, it has also been true that CANDUs have suffered few of the operating incidents or operator errors that have led to the erosion of public confidence in the USA. The incidents that have occurred seem to have been handled competently and, unlike the Browns Ferry fire and the TMI accident, have not been seen as 'near misses'
- CANDU has enjoyed a high public status as one of rather few successful indigenous ventures in high technology
- CANDU's renunciation of any involvement in nuclear weapons has helped remove one potent source of suspicion and hostility to civilian nuclear programmes. Canada has needed no major enrichment plants and has no plans to reprocess spent fuel

OVERALL EVALUATION

The Canadian CANDU programme must be counted a remarkable technical achievement. With no international experience to draw on, a relatively small nation with no great traditions in this area has produced a reactor which has consistently out-performed every other system in the world. It is axiomatic that any successful programme of nuclear power reactors must embody the following features:
- a good basic design concept
- a good detailed station design
- high quality components and construction work
- well-trained operators and sound operating procedures
- high quality maintenance

The first, and to a lesser extent, the second preconditions are difficult to plan for and cannot be achieved simply by good, tight engineering practice and policy. Given the existence of these two preconditions, Canada's success lies in the way that the three key institutions Ontario Hydro, AECL and AECB have ensured that sufficient attention is paid to the other preconditions.

A number of features of the achievement deserve more detailed discussion since in some cases they run counter to a number of conventional wisdoms; in particular the views that standardisation is an essential element of a nuclear programme; that increased scale necessarily brings economies; and that as complex a technology as a nuclear plant requires resources only available to a widely diffused technology to ensure good operation and problem solving.

Standardisation

AECL/Ontario Hydro have not placed undue emphasis on standardisation. There have been four basic designs, two of which are still offered – the superseded ones are the Douglas Point/Rajasthan 200 MW design and the Pickering 500 MW design, whilst current designs are the Bruce/Darlington 900 MW design, and the Point Lepreau/Gentilly 600 MW design. Even within these basic models there have been important design and material changes. These include:
- the change of pressure tube material half-way through the first tranche of Pickering units
- the differing materials used for steam generators
- the change from integral to non-integral steam drums and steam generators from Pickering to Bruce
- the greater allowance for pressure tube creep mid-way through the first Bruce tranche

Most of these have had important impacts on the operation of plant.

This lack of standardisation may have brought cost penalties compared to a strategy of series ordering of a single standard design (had this been feasible). However the 'first-off' nature of components, together with the policy of maintaining competition amongst suppliers may have stimulated manufacturers to pay particular attention to quality in order to put themselves in a good position for subsequent orders. Such a policy is not risk-free as Babcock & Wilcox's errors in steam generator manufacture illustrated. When the likelihood of subsequent orders is low, this policy is difficult to maintain.

Manufacturing of components has not been problem-free and significant errors have occurred, notably the faults in the pressure tubes at Pickering 3 and 4 (which came to light in 1974), and the incorrect heat treatment applied to the steam generators to be used at Pickering B, Gentilly 2, Point Lepreau and Embalse. In addition, the error in the design of the garter springs, a relatively simple piece of equipment, was a serious one. Had the Canadian programme of building proceeded at somewhere near the same rate as the French programme, these errors would have had serious economic consequences for Ontario Hydro and for the future of the technology as a whole. The much slower rate of building and the consequent absence of dependence by Ontario Hydro on a particular vintage of plant together with early identification of potential problems has allowed them to evolve concerted strategies to produce long-term solutions to these problems.

More generally, whilst it is difficult to make meaningful international comparisons, the low failure rates experienced suggest that the quality of relatively small common components such as valves, pipe-work, pumps etc., has been exceptionally high. In addition the steam generators have been very much more trouble-free than US designs.

Scale economies

A major difference between the approach of AECL/Ontario Hydro to commercialisation and that of other manufacturers lies in the extent to which economies of scale have been sought. Manufacturers in France, FR Germany and the USA have moved to the largest size licensable in the USA – 1300 MW – and the French are now going beyond this size. This seems to have been due to a belief that larger reactor sizes would reduce unit costs and perhaps give the vendor an edge in world markets. In Canada where technology development was much more strongly influenced by the customer (Ontario Hydro) than it was elsewhere, the view on scale economies was different. Feasibility studies on larger reactors were carried out but were not pursued beyond this stage. There were two main reasons for this decision:

– AECL and Ontario Hydro were sceptical of the economies of scale to be gained by larger reactors. They anticipated that 1300 MW reactors would achieve poorer operating performance, that the scale economies of building larger could be matched by the volume economies of building more units, and the latter strategy offered more operational flexibility. These were reinforced by the greater difficulties in financing large reactors

– the markets in which the CANDU had the best prospects were not well suited to the largest size of reactors

International technological support

The major area where the technology may have suffered is in pressure tube technology. It is possible that a more diverse set of customers and perhaps a foreign licensee would have identified the factors that led up to the pressure tube rupture especially the error with the garter spring. The tube material which was susceptible to rupture was superseded at a very early stage (after the first two commercial units) and it is unlikely that international diffusion would have helped in this area. However, in this, as in other problem areas, Ontario Hydro and AECL alone seem to have been able to arrive at solutions with relatively little delay. This may partly be a reflection of the relatively less complex nature of CANDU technology.

Overall the lack of a Canadian tradition and infra-structure in the heavy electrical sector may paradoxically have had advantages. Unlike France and Britain, Canada has not had to struggle hard to re-shape its industries to meet the new requirements. The structure that has evolved has proved a good one.

A final, less tangible, factor that has contributed to the success of CANDU lies in the Canadian technological environment. The prestige nature of the technology and the relative scarcity of ventures in Canada offering a comparable challenge has meant that high quality human resources have been attracted to the programme. Further, the lack of development of alternative reactor systems has meant that their effort has been concentrated rather than thinly spread.

FUTURE PROSPECTS

The challenges that the Canadian nuclear industry will face over the next ten years are at least as significant as those it has met over the past 20 years. At the same time, many of the conditions which have contributed to past success will no longer be present.

On the industrial side, as for all suppliers, the difficulties will be to maintain critical resources in design and manufacturing in a period when demand will be low and customers will have little appetite for new and untried designs. AECL, being software-based has some advantages in this, as changing regulatory requirements and reactor maintenance and repair may provide a significant flow of work requiring their skills. On the hardware side, AECL itself will not sustain direct losses, but its policy of 'dual sourcing' components will probably not be feasible. Costs are also likely to increase as overheads have to be spread over fewer orders and especially if components have to be bought on a 'one-off' basis. Export orders, even if possible, may not alleviate this situation much, as potential customers such as South Korea, Argentina and Romania will aim to maximise domestic content by manufacturing as many components as possible. In addition Ontario Hydro is playing an increasingly independent role in design and technology development. This is likely to increase the difficulties facing AECL in maintaining resources, unless some agreement can be reached between AECL and Ontario Hydro to pool resources.

The bid put forward by AECL to build and operate a plant in Turkey selling its electrical output to the Turkish electric utility is a bold but risky bid to break the impasse in finance that seems to be blocking export sales for all vendors. The risks are well illustrated by the experience of AECL with Gentilly 1 where AECL lost considerable sums of money because of the poor performance of this unit and Hydro-Quebec's understandable refusal to buy the plant. With the Turkish bid, the technology is at least not untried although it may turn out to be more difficult to control site-work quality in Turkey than Quebec.

The pressure tube rupture has put some question-marks against the technology itself. The theory that the pressure tubes would leak detectably before rupture no longer seems valid although since the rupture caused little damage this may not be a very significant change. The best outcome will occur if the pressure tube replacement can be completed to schedule, the units operate well, and it is firmly established that blistering is not likely to be a problem with the later units. This outcome will open up the prospect of a very much longer life for CANDUs and will also give AECL/Ontario Hydro experience in carrying out repairs to highly contaminated areas which may open up a useful stream of international business not confined merely to CANDU reactors.

The problem that the Canadian industry faces more acutely than those of the other three countries covered in this book, is that its domestic market for reactors is particularly constrained for the foreseeable future. An export sale to a prestige market such as Italy, the UK or Japan would radically alter CANDU's prospects but the chances of this seem remote. Even more remote, for political and regulatory reasons, is the prospect of sales to the closest foreign market of the USA. But export orders will do little for hardware sales in the long term (because of the desire of importers to 'indigenise' the technology) unless a whole series of new markets could be successively opened. This is a daunting prospect for any industry.

A particular difficulty may concern the performance and safety record of CANDUs away from their Ontario base and the undoubted skills of Ontario Hydro. The record and reputation of CANDUs has been built on performance at only two sites, and much will depend on the capacity of other utilities – both in Canada and outside – to reproduce the significant achievements of Ontario Hydro.

APPENDIX. DETAILED ANALYSIS OF PLANT PERFORMANCE OF
CANADIAN CANDU PLANT

Table A7.1. *Causes of lost output in mature CANDU units according to time period (%)*

Time period	No. of reactor years	Capacity factor	Shutdown losses	Derating losses	Operating losses	Adjusted operating losses
Pickering						
1974–78	17	84.8	13.4	0.1	1.6	1.9
1979–84	24	76.6	22.3	0	1.1	1.2
Bruce						
1979–84	17	92.7	8.1	0	−0.7	−0.7

Source: SPRU nuclear power plant data base.

Table A7.2. *Causes of lost output in CANDU units according to age (%)*

Age	No. of reactor years	Capacity factor	Shutdown losses	Derating losses	Operating losses	Adjusted operating losses
Pickering						
1	3	85.8	13.7	0.1	1.0	1.1
2	4	57.8	39.4	0.2	2.6	5.7
3 onwards	41	80.0	18.6	0.1	1.3	1.5
Bruce						
1	4	67.4	24.7	0	7.9	10.8
2	4	81.4	15.0	0	3.6	4.5
3 onwards	17	92.7	8.1	0	−0.7	−0.7
Gentilly/Point Lepreau						
1	2	75.4	16.8	−0.5	8.3	10.8

Source: SPRU nuclear power plant data base.

Table A7.3. *Causes of shutdown in CANDU units according to age*

Age	No. of reactor years	Shutdowns (hours per year of operation)													Shutdown frequency (no. per year of operation)
		1	3	4	5	6	7	8	9	10	B	D	E	H	
Pickering															
1	3	0	64	529	77	0	0	31	0	11	29	218	101	185	6.2
2	4	0	37	0	2230	0	31	1212	0	0	23	334	500	334	8.1
3 onwards	41	528	24	24	205	11	8	233	14	21	19	583	0	5	4.9
Bruce															
1	4	0	122	138	143	302	0	419	34	0	79	292	974	133	13.7
2	4	0	136	245	0	242	18	327	46	16	0	487	0	10	8.6
3 onwards	17	45	35	38	179	49	6	76	22	48	3	214	8	12	4.8
Gentilly/Point Lepreau															
1	2	0	0	43	0	0	3	385	0	0	47	1129	2	0	8.2

Causes of shutdown are:
1 = reactor and accessories
3 = reactor control system and instrumentation
4 = nuclear auxiliary and emergency system
5 = main heat removal system
6 = steam generators
7 = feedwater, condenser and circulating water system
8 = turbine generator
9 = electrical power and supply system
10 = miscellaneous equipment failure
B = operating error
D = maintenance and repair
E = testing of plant systems/components
H = other

Source: SPRU nuclear power plant data base.

Table A7.4. *Causes of shutdown in mature CANDU units according to time period*

Time period	No. of reactor years	Shutdowns (hours per year of operation)													Shutdown frequency (no. per year of operation)
		1	3	4	5	6	7	8	9	10	B	D	E	H	
Pickering															
1974–78	17	4	22	35	350	15	9	253	25	46	22	493	0	13	4.7
1979–84	24	991	25	16	102	7	7	219	7	3	16	646	0	0	5.0
Bruce															
1979–84	17	45	35	38	179	49	6	76	22	48	3	214	8	12	4.8

Note: See Table A7.3 for causes of shutdown.
Source: SPRU nuclear power plant data base.

CHAPTER 8

France

Introduction

In all of the countries which made an early commitment to nuclear power, strategic factors such as reducing external vulnerability, insuring against fossil-fuel shortage, and satisfying military needs were of key importance. As the contribution of nuclear power developed, economic factors generally assumed much greater importance. However, in France strategic considerations have maintained their important influence on decision-making. The massive scale of the post-1973 French programme has meant that the future prosperity of France will be significantly affected by the degree of technical, economic and export success of the technology.

In this chapter we examine how successful the largest attempt at thorough-going standardisation of power station design has been, and the way in which central governmental control has allowed a vast programme of reactors to be completed to a tighter time and budgetary schedule than anywhere else in the world. We also examine how the system may respond to the strains that will result if, as now seems inevitable, electricity demand does not meet expectations, the burden of the debt repayment continues to cause problems to the utility (Electricité de France), and the manufacturer (Framatome) fails to win a sufficient volume of export orders to compensate for a decline in home orders.

THE ENERGY AND ECONOMIC BACKGROUND

Since 1970 the French economy has followed a path which is fairly typical amongst Western industrialised countries. Prior to the first oil crisis GDP growth was over 5% per annum. In the period 1974–79 growth fell to about 3% and, following the Iranian revolution and the subsequent world recession, fell to below 1% until the recovery of 1984.

The energy data (see Table 8.1 and Table 8.2) reflect the very low reserves of indigenous resources and the effects of the policies to diversify

Table 8.1. *Production and consumption of primary energy in France – 1973–84 (million tonnes of oil equivalent)*

	Production			Consumption			Consumption for electricity generation		
	1973	1978	1984	1973	1978	1984	1973	1978	1984
Solid fuels	19.0	15.5	12.4	30.8	32.9	26.1	10.7	17.6	13.3
Liquid fuels	1.3	2.0	2.8	122.6	115.0	85.5	15.6	11.8	1.8
Natural gas	6.4	6.7	5.4	13.9	19.6	23.9	2.1	1.4	0.7
Nuclear power	3.3	6.8	42.7	3.3	6.8	42.7	3.3	6.8	42.7
Other primary electricity	10.8	15.5	15.2	10.8	15.5	15.2	10.8	15.5	15.2
Total	40.8	46.5	78.5	181.4	189.8	193.4	42.5	53.1	73.7

Source: Energy balances of OECD countries, 1970–1982 and 1983–1984, International Energy Agency, Paris, 1984 and 1986.

Table 8.2. *Delivered energy consumption in France – 1973–84 (million tonnes of oil equivalent)*

	Industry			Transportation			Residential[a]			Other[a]			Total		
	1973	1978	1984	1973	1978	1984	1973	1978	1984	1973	1978	1984	1973	1978	1984
Solid fuels	12.3	9.3	8.6	0.1	–	–	6.0	4.1	2.6	–	–	–	18.4	13.4	11.2
Liquid fuels	27.3	26.5	17.0	26.4	30.7	33.8	32.5	26.1	2.2	5.2	6.8	20.4	91.4	90.1	73.4
Gas	5.7	9.0	10.9	–	–	–	3.6	4.9	6.6	1.9	3.7	5.4	11.2	17.6	22.9
Electricity	7.7	9.6	10.2	0.6	0.6	0.7	2.6	4.6	6.9	2.4	2.2	3.0	13.3	17.0	20.8
Total	53.0	54.4	46.7	27.1	31.3	34.5	44.7	39.7	18.3	9.5	12.7	28.8	134.3	138.1	128.3

[a]The allocation of energy demands between the 'residential' and 'other' sectors does not appear to be consistent or satisfactory.
Source: See Table 8.1.

away from imported energy, particularly oil, resources. As seen in 1974, indigenous resources of oil and gas were very limited, coal was expensive to produce by world standards and the potential to expand hydroelectric production was limited. This left only nuclear power and conservation as options for reducing energy import dependence. Further, since oil for electricity generation represented only about 13% of total oil use, a substantial reduction in oil consumption would require the substitution of (nuclear) electricity for the direct use of oil. Given the long lead-times involved in building nuclear power stations and persuading consumers to replace their energy-consuming equipment by electrically-powered equipment, this policy's full effects have still to be felt. Nevertheless by 1985 total consumption of electricity in France was almost equal to the sum of production from nuclear and hydro resources* and electricity's market share of final consumption was almost double its 1973 level.

On the demand side, it can be seen that electricity's progress has been particularly strong in the residential sector and also in the industrial sector where it has increased its market share by 50%.

Improved energy efficiency also appears to have been an important factor with primary energy use per unit of GDP 18% lower in 1984 than it was in 1973 – about the same percentage reduction as the rest of OECD Europe.†

THE DEVELOPMENT OF NUCLEAR POWER

The development of civil nuclear power can best be seen in four phases; these correspond almost exactly, and not by coincidence, with the periods of office of the four Presidents of the Fifth Republic.

In the first phase up to 1969, the de Gaulle era, indigenous French reactor designs were predominant, notably the GCR and the FBR. The first civil reactors (GCRs) entered service or were nearing completion and an export order (to Spain) was won. However, it became apparent that without significant development work, the GCRs could not compete with American LWRs. Whilst the Commissariat de l'Energie Atomique (CEA), the state nuclear research organisation, emerged as a powerful and politically influential organisation (partly through the prestige associated with its military activities), championing the cause of French technology, no strong

---

* Because of the 'peakiness' of electricity demand fossil-fired stations are still required to meet peak demands but a considerable quantity of electricity is available for export at non-peak times.

† Reductions in energy use per unit of GDP should be interpreted with care as they take no account of structural change in the economy which may reduce overall energy intensity with no real improvement in energy efficiency.

reactor supply industry had emerged. Electricité de France (EdF), the national state-owned electric utility strongly backed the adoption of American LWR technology.

The struggle was resolved at the start of the second phase (the Pompidou era, 1969–75) by the decision to build both BWRs and PWRs under licence to General Electric and Westinghouse. The plant supply industry was to be strengthened through a series of restructuring deals. The 1973–74 oil crisis substantially increased the pace of development and in 1975 a further series of deals allowed concentration exclusively on the PWR and the ordering of a series of highly standardised reactors.

During the third phase (1976–81, the Giscard era), resources were concentrated on carrying through the programme which comprised about six orders a year. Developments to the technology and design changes were minimised until resources had been built up further and also to avoid the risk of building a long series of reactors, significant elements of which were unproven.

By the 1981 elections, concern had built up about the autocratic nature of the procedures, the apparent over-investment in plant and the concentration of resources on nuclear development to the exclusion of other options. The subsequent Mitterand era (1981 onwards) has been one of adjustment and consolidation. The pace of ordering has slowed although the momentum of the nuclear programme has proved formidable. Framatome (the plant vendor) despite considerable organisational and ownership changes, has remained a strong force and has placed greater emphasis on reactor development and establishing itself in world markets.

THE KEY INSTITUTIONS

The institutional structure of nuclear power in France is more tightly knit, closed to external influence and centrally directed than that of any other Western user of nuclear power. The three key actors are, EdF, Framatome and the CEA. This latter organisation, aside from its basic R&D role, has a wide involvement in regulation, all aspects of the fuel cycle and a financial stake in other important companies. The Ministry of Industry held a key decision-making and co-ordinating role in the period from 1974–81 when the current situation was largely shaped.

Electricité de France (EdF)

EdF was nationalised in 1946 (along with the national coal, oil and gas companies) and it supplies virtually all electricity sold to final consumers. About 90% of its sales are supplied from its own plant and most of the balance is purchased from Charbonnage de France (CdF), the state coal company, and Compagnie National du Rhône, a company with responsibility for developing hydro-electric resources in the south.

In terms of capacity of plant owned, EdF is now the largest electric utility in the world and has a commensurately large engineering capability. It also owns about four times as much nuclear capacity as any other utility in the world. It now has a 10% holding in Framatome and participates in export orders with Framatome, offering its nuclear expertise.

The plant supply industry

The plant supply industry is based around the now largely state-owned company Framatome which supplies the nuclear island and Alsthom, a subsidiary of the state-owned Compagnie Generale d'Electricité (CGE). There have been frequent changes in the ownership and scope of business of these companies (covered in more detail in the following sections) as they have expanded to meet the needs of the large ordering programme of the past 10 years.

Framatome's ownership pattern is currently 40% by CGE, 35% by CEA, 12% by Dumez, a public works group, 10% by EdF and 3% by Framatome management. Despite the prolonged power struggle and upheavals which have continued since 1981, Framatome appears to have been little affected in terms of its technical and commercial effectiveness. A feature of Framatome is its high degree of horizontal integration and the very sophisticated production techniques used.

Two points are worth noting about both Framatome and CGE. First, both are currently firmly under state control and second, neither company has a long tradition at the forefront of innovation and technology leadership in the heavy electrical industry in the way that some German and American companies do. Neither do they have well-established export markets where they can be sure of a reasonable volume of orders.

Government

There is a long history of centralised state-control and planning in France with industrial development strategies being carried through with considerable determination and constancy. As was mentioned above, the key ministry in the expansion period of the late 1970s was the Ministry of Industry, at which point most key decisions were ultimately taken. However, this may have had more to do with the personality and drive of the then Minister, Andre Giraud (previously head of the CEA). Since the 1981 elections, the Ministries of Planning, Energy and Technology have played a greater role.

The Ministry of Finance has always been thought to have been powerful but it appears to have had little success (or has not tried) in restraining public expenditure on nuclear power, the traditional stance of treasury departments.

Commissariat a l'Energie Atomique (CEA)

As with other pioneers of nuclear power such as the USA and the UK, the French government created a national body, the CEA (in 1945) with wide responsibilities for developing all aspects of nuclear power including both civil and military applications. As with the AEA in the UK and the AEC in the USA, these responsibilities have changed through time. However, the CEA has retained more of its early responsibilities and interests, notably in reactor design (through its stakes in Framatome and the FBR vendor, Novatome) and safety regulation than its counterparts. Also the development of the fuel cycle, which has been more comprehensive and on a larger scale than anywhere else in the world, has remained under its control through wholly owned subsidiaries such as Compagnie Generale des Matières Nucleaires (COGEMA).

Effectively the CEA has a very strong say in every aspect of civil nuclear power from uranium exploration (COGEMA), through reactor design and manufacture (Framatome) to safety regulation.

Safety regulation

The formal regulatory body is a part of the Ministry of Industry, the Service Central de Sûreté des Installations Nucleaire (SCSIN). This is a small, largely administrative body advised by a division of the CEA, the much larger Institut de Protection de Sûreté Nucleaire (IPSN). The IPSN, which currently has a staff of about 1000, is responsible for evaluating new proposals, ensuring quality control during construction, licensing for operation and dealing with operating problems as they arise.

The licensing procedure was tailored to allow a large programme of reactors to be completed to time and so an important factor was that opportunities for intervenors to cause the sort of delays either by direct or legal means that had slowed the American and German programmes should be minimised. The licensing process is simple, the requirements are specified in a very general way and public hearings are strictly limited.

The formal stages of the licensing procedure are as follows:
– the applicant (EdF) sends a summary specification of the installation

including a preliminary safety report. That is scrutinised by all the relevant ministries – Interior, Health, Equipment and Region Development, Agriculture, Environment and Transport and a public inquiry is set up
- at the public inquiry, often held before the licence application is made, details on architectural aspects of the installation, environmental impact and the main measures for nuclear safety and radiation protection are outlined. This procedure results in a Declaration d'Utilite Publique (DUP)
- concurrently the SCSIN in consultation with the IPSN evaluates the safety report and, provided its verdict and those of the various ministries are favourable, a draft decree is sent to an interministerial committee for a final decision on whether construction can begin. The decree specifies the requirements that the installation must meet in terms of safety, environment etc.
- six months prior to loading of fuel, EdF must send to the Ministry of Industry a provisional safety report and the operating instructions. If the Minister, in consultation with the competent bodies finds these satisfactory, approval is given for fuel loading and commissioning tests
- whilst testing is occurring, a final safety report is submitted and, if approved, the Minister of Industry will issue a full operating licence

Fuel cycle facilities

Fuel cycle development has played a much more central role in French nuclear power development than in other countries. The imperative during the PWR expansion period of the 1970s clearly lay with developing the front end of the fuel cycle, especially the capital-intensive enrichment stage, to ensure an adequate supply of fuel for the reactors. Nevertheless the back-end of the fuel cycle, reprocessing, waste disposal, etc., has received considerable attention. The key companies in this are COGEMA, a wholly owned subsidiary of the CEA and Pechiney Ugine Kuhlman (PUK). COGEMA was set up in 1975 with private legal status which helped it in creating joint ventures. It controls the enrichment and reprocessing aspects of the fuel cycle whilst PUK is concerned with fuel fabrication.

Prior to 1979, enriched uranium was imported from the USA and the USSR. Since 1979, the multi-national Eurodif plant (sited at Tricastin along with four 900 MW PWRs) has begun to take over the workload. Eurodif, designed to have a capacity of 10.7 million tonnes separative work units (SWU) per annum was originally to provide enriched uranium to France, Belgium, Spain, Italy and Iran, with France's shareholding the largest at 42.8%. In fact, as the other countries' nuclear programmes have fallen short of expectation, France has taken an increasing share of production and the necessity for a second enrichment project, Coredif, has receded far into the future.

THE DE GAULLE ERA (BEFORE 1969)

The key force in the early development of nuclear power was the CEA. The motives behind its creation were:
- to establish a position for France as a major force in nuclear technologies
- to increase France's independence from imported fuel supplies and, more generally, from foreign technology
- to help re-establish France as a world power through re-assertion of its military strength by means of nuclear weapons

This third motive was initially less overt than in the UK and the USA. However, the first power reactors at Marcoule were primarily designed to produce plutonium for use in France's test explosion programme which started in 1960. One consequence of military activity for the civil programme was that the CEA gained early experience in PWR technology through its independent development of submarine reactor designs.

As in all Western countries outside the USA, the US ban on the supply of both enriched uranium and enrichment technology initially constrained France to natural uranium reactors which had to be moderated by graphite or heavy water. The reactors were ordered in three main 'tranches' (see Table 8.3). The first was started in 1956 and consisted of three units at Chinon. In 1958 Framatome took out a licence from Westinghouse for PWR technology, and in 1960 EdF decided to diversify reactor technologies and ordered a PWR, Chooz, in a joint venture with a Belgian company. The second tranche of gas/graphite reactors was started in 1963, comprising St Laurent 1 & 2, Bugey 1 and a joint venture in Spain at the Vandellos site. This latter unit is 25% owned by EdF and is similar in design to St Laurent 2.

In addition to these efforts, which were led by EdF, the CEA was developing other reactor concepts, notably the FBR, through the Rapsodie reactor and a heavy water reactor through the Monts d'Arree project.

By the mid-1960s there was pressure from the French heavy electrical industry and EdF, stimulated by the watershed Oyster Creek order in the USA, to switch out of GCR technology into LWRs. The Oyster Creek bid, the first time in the USA that a nuclear bid appeared to beat all conventional bids purely on economic grounds, gave expectations of considerable returns to scale for LWRs (a feature not expected for GCRs) and thus a widening cost advantage for LWRs over GCRs. The CEA argued strongly that the strategic considerations – mainly technological independence – were not

Table 8.3. *The French nuclear power programme[a] – orders from 1956–68*

Name	Type[b]	MW (gross)	Start of construction	Year of commissioning	Year of decommissioning	NSSS supplier[c]
Marcoule G2	Gas/graphite	42	1956	1959	1980	SACM
Marcoule G3	Gas/graphite	42	1956	1960	1984	SACM
Chinon 1	Gas/graphite	83	1957	1964	1972	Various
Chinon 2	Gas/graphite	242	1958	1966	1985	Various
Chinon 3	Gas/graphite	500	1961	1967	–	Various
Chooz[d]	PWR	325	1962	1970	–	ACECO/Fra
Monts d'Arree	GCHWR	75	1962	1967	1985	CEA/Indatom
St Laurent 1	Gas/graphite	500	1963	1969	–	Various
Bugey	Gas/graphite	560	1965	1972	–	Various
St Laurent 2	Gas/graphite	530	1966	1971	–	Various
Phénix	FBR	250	1968	1974	–	CEA/EdF/GAAA

[a] Includes all reactors with electrical output over 30 MW.

[b] Reactor types:

Gas/graphite – carbon dioxide cooled, graphite moderated, natural uranium fuel
PWR – Pressurised water reactor, light water cooled and moderated, enriched uranium fuel
GCHWR – carbon dioxide cooled, heavy water moderated, enriched uranium fuel
FBR – fast reactor, sodium cooled, plutonium/uranium fuel

[c] NSSS suppliers:

SACM – Société Alsacienne de Construction Mecanique
ACECO – Ateliers de Construction Electrique de Charleroi/Cockerill Ougree Providence (Belgium)
Fra – Framatome
CEA – Commissariat a l'Energie Atomique
EdF – Electricité de France
GAAA – Groupement Atomique Alsacienne Atlantique

[d] This unit is jointly owned 50% by EdF and 50% by SENA (Belgium).

Source: SPRU nuclear power plant data base.

given sufficient weight in this view and advocated the development of high-temperature gas-cooled reactors and FBRs. A consultative commission, the Commission Consultative pour la Production d'Electricité d'Origine Nucleaire (PEON) was instructed to advise the government on reactor choice. Their report, completed in April 1968 and published in December 1968 came down largely in favour of LWRs at the expense of GCRs. However, it was not until after de Gaulle, a powerful supporter of indigenous French technology, left office in April 1969 that the government acted on this report. The fundamental reorganisation and redirection of French industry that took place in the 'Pompidou' era is discussed in the next section.

THE POMPIDOU ERA (1969–75)

The main feature of this period was the evolution of a large ordering programme and the changes and rationalisation in the supply industry that followed the death of de Gaulle. Under the Pompidou administration, the recommendations of the PEON Commission were broadly adopted and over the next six years, there followed a complicated series of mergers and rationalisations which are summarised in Table 8.4. (For further details of the reorganisation and other features of this period see Torresi (1972, pp. 702–3).)

In November 1969, following the report of the PEON Commission, the French government effectively abandoned GCR development in favour of American LWR technology. Some work was to be continued on heavy water reactors and high-temperature reactors but the main alternative strand of reactor development was to be the FBR. At that time Framatome was only a design and management company and had just commenced work on an 870 MW PWR in Belgium (Tihange) in collaboration with a Belgian company (ACECO, later taken over by Westinghouse) via a licence with Westinghouse. The design for this unit differed little from the then current American designs. In March 1970, the Société des Forges et Ateliers du Creusot (SFAC) which owned the Framatome company, and the Compagnie General d'Electricité (CGE) were invited to tender for a 900 MW station at Fessenheim and it was left to the companies to decide whether they should offer a BWR or a PWR. In fact SFAC submitted a bid for a PWR whilst CGE–Alsthom entered a bid for a BWR via a licence from General Electric (USA). Ultimately all six units in the Programmes '1970' tranche ordered between 1970 and 1974 were won by Framatome, with Alsthom winning the turbine orders. Much of this period was spent in attempts to rationalise the French heavy electrical business. This was a part of a general industrial strategy initiated by Pompidou, the main purpose of

Table 8.4. *Key decisions in the reactor supply industry 1969–76*

November 1969	Decision to abandon GCR development and concentrate on American LWRs.
March 1979	Call for tenders for building an LWR at Fessenheim.
October 1970	SFAC (Creusot) wins Fessenheim order.
June 1971	Babcock Atlantique announces a capability in PWRs through links with Babcock & Wilcox (USA).
October 1971	SFAC wins next tranche of LWR orders with two PWRs for Bugey (subsequently increased to four units).
November 1972	Re-formation of Framatome with majority shareholding to Creusot Loire.
July 1973	Announcement of plans to order two BWRs from SOGERCA at St Laurent des Eaux.
January 1974	Firming of order for two BWRs and option taken for six further units.
March 1974	Ordering of 12 PWRs.
March 1974	Negotiation of sale of 5000 MW of PWRs to Iran in exchange for oil advantages.
July 1974	Formal ordering of Super Phénix.
August 1975	Abandonment of BWR option.
August 1975	CEA takes 30% holding in Framatome.
April 1976	Novatome created to advance FBR development. Ownership divided between Creusot-Loire, Alsthom and CEA.

which was to concentrate the main sectors of French industry into large and sometimes monopolistic units able to compete effectively in international markets.

The two key industrial forces were the Empain–Schneider group and the CGE group. The Empain–Schneider group had acquired interests in conventional and nuclear boilers (through Creusot–Loire), PWRs (through Framatome) and turbine generators (through Jeumont Schneider and Merlin–Gerin). CGE's interests were in BWR supply (through SOGERCA), conventional and nuclear boilers (through Stein Industrie) and turbine generators (through Alsthom, Rateau and Neyrpic). A third group, Babcock–Atlantique, was also competing in these areas through the supply of Babcock & Wilcox designed PWRs, but never attained a strong position of influence. The other important company was the French subsidiary of Brown Boveri (Switzerland), the Compagnie Electro-Mecanique (CEM) which had a strong position in turbine generator supply.

The first fruit of this process was the re-formation of Framatome in 1972 with a major holding going to Creusot–Loire (51%) and most of the rest of the holding going to Westinghouse (45%). Through the period up to 1975, there was a strong expectation of orders for BWRs and in 1973 EdF issued a letter of intent to CGE which began to make preparations through internal rationalisation and international collaborations (e.g. with Asea (Sweden) and KWU (FR Germany)).

Turning to the second main topic of interest in this period – the build-up of nuclear ordering – the first important point to note is that the ordering programme had been building steadily from 1970, well in advance of the oil crisis of 1973–74. In the 1968–70 period, EdF had evolved, under Boiteux, an aggressive marketing strategy. (See Lucas, 1977, pp. 139–40.) This was designed to restore flagging electricity demand growth and a central element was the promotion of an LWR-based nuclear strategy. Thus in the sixth French plan (1971–75) there was a target to order 8000 MW of LWR capacity. When the PEON Commission met again in 1972 it recommended that 13 000 MW of LWR capacity should be ordered in the five years, 1973–77 (Lucas, 1979, p. 148). (Table 8.5 shows the actual pattern of orders.) This strong commitment, a full year in advance of the oil shock, was largely due to official French sensitivity to external energy dependence in the wake of the Algerian nationalisation of French oil interests, and the Teheran–Tripoli agreements of 1970–71. Hence when the crisis struck in 1973, there was already significant momentum in the nuclear expansion. In late 1973, PEON recommended that 13 900 MW units be started in 1974 and 1975, and this was accepted by the government. On the export front, access to Iranian oil was negotiated using the export of 5000 MW of PWRs as a lever.

Table 8.5. *The French nuclear power programme – orders from 1970 to 1986*[a]

Name	Tranche	MWw (net)	Year of order	Year of commissioning
Fessenheim 1, 2	Programmes '1970'	2 × 920	1970, 71	1977, 78
Bugey 2, 3, 4, 5	Programmes '1970'	2 × 955, 2 × 937	1972, 72, 73, 74	1979
Tricastin 1, 2, 3, 4	CP900–1	4 × 955	1974, 74, 75, 76	1980, 80, 81, 81
Gravelines B 1, 2, 3, 4	CP900–1	4 × 951	1974, 75, 75, 76	1980, 80, 81, 81
Dampierre 1, 2, 3, 4	CP900–1	4 × 937	1974, 75, 75, 76	1980, 81, 81, 81
St Laurent B 1, 2	CP900–1	2 × 921	1976	1981
Le Blayais 1, 2	CP900–1	2 × 951	1976, 77	1981, 82
Chinon B 1, 2	CP900–2	2 × 919	1977	1984
Le Blayais 3, 4	CP900–2	2 × 951	1977, 78	1983
Cruas 1, 2, 3, 4	CP900–2	4 × 921	1978, 78, 79, 79	1984, 85, 84, 85
Paluel 1, 2, 3, 4	CP1300–1 (P4)	4 × 1344	1977, 77, 78, 80	1985, 85, 86, 86
St Maurice/St Albans 1, 2	CP1300–1 (P4)	2 × 1348	1979, 80	1986, 87
Flamanville 1, 2	CP1300–1 (P4)	2 × 1344	1979, 80	1986, 87
Creys Malville (Super Phenix)	–	1240	1976	1987
Gravelines C 5, 6	Tranches 900	2 × 951	1979, 80	1985
Chinon B 3, 4	Tranches 900	2 × 919	1981, 82	1987, 89
Cattenom 1, 2, 3, 4	Tranches 1300 (P'4)	4 × 1330	1979, 80, 82, 84	1987, 87, 90, 93
Belleville 1, 2	Tranches 1300 (P'4)	2 × 1330	1981	1988, 90
Nogent 1, 2	Tranches 1300 (P'4)	2 × 1330	1981, 82	1989, 90
Penly 1, 2	P'4, N4	1344, 1450	1981, 82	1991, 94
Golfech 1, 2	P'4, N4	1330, 1450	1983, 85	1991, 95
Chooz B 1	N4	1450	1984	1992

[a]With the exception of Creys Malville, all reactors are PWRs with NSSS supplied by Framatome, turbine supplied by Alsthom or CEM and operated by EdF. Creys Malville is a multinational project, 51% owned by EdF with the rest owned by ENEL (Italy 33%) and RWE (FR Germany, 16%). The NSSS is supplied by Novatome and the turbine generators by Ansaldo (Italy).

Source: SPRU nuclear power plant data base.

By 1975, however, it was clear that things were not proceeding fully as planned and 2 BWR orders were postponed in May. At the same time, CEA was negotiating a stake in Framatome, which implied government endorsement of a PWR-only strategy. In August the deal was settled, the BWR abandoned and the CEA purchased a 30% share in Framatome from Westinghouse with an option to buy the rest of the Westinghouse holding in 1982. The anticipated scale of the French programme meant that nearly all interests could do well from the deal. The Empain–Schneider camp was given a monopoly in reactor supply through Framatome whilst the CGE–Alsthom group were given two-thirds of the turbine generator orders. The remainder of the turbine generator orders went to CEM. CEM had been expected to supply all the turbine generators for the BWRs and, through Brown Boveri, had access to proven technology for 1300 MW units to which EdF were planning to move. From this point, PWRs were to be ordered at five to six a year for the rest of the decade in a series of five tranches.

Over the same five- to six-year period, the CEA was attempting to re-establish an important role following the rejection of its gas-cooled and heavy water technologies. Their influential chairman, André Giraud, played an important role in this. The massive regulatory and fuel cycle commitments which resulted from the new programme fell fairly naturally on the CEA. A new daughter company created in 1976, COGEMA, handled the fuel cycle business, notably the Eurodif project which provided enriched uranium and the reprocessing facilities at La Hague. In the FBR field, the CEA subsidiary Technicatome had been competing with an off-shoot of CGE–Alsthom, the GAAA, for FBR work but, in 1976, a new company, Novatome (40% Creusot–Loire, 30% Alsthom, 30% CEA), was created to consolidate this expertise. These developments, along with a key role in PWR safety research and a substantial holding in the nuclear business, restored the CEA to an important position in the nuclear establishment. The changes also removed the frictions between the CEA and EdF that had been apparent in the later 1960s over reactor choice.

THE GISCARD ERA (1974–81)

With the decision in August 1975 to abandon the BWR, a decision long anticipated and probably of more formal than actual significance at that point, the way was clear to seal important 'deals' in the heavy electrical sector and to initiate a major programme of nuclear orders.

The programme was to be carried through with a tight administrative arrangement involving the CEA, EdF and Framatome with the Ministry of Industry at the centre of the triangle performing a co-ordinating and channelling role. Foremost among the three, however, was EdF. Once

LWRs had replaced GCRs, EdF assumed the primacy in the French nuclear establishment that the CEA had previously occupied. It was subsequently the ambitions of EdF, first formulated in the late 1960s, that largely drove the expansion in the 1970s. EdF was closely supported in such important and influential bodies as PEON both by the CEA and Framatome (as well as other private interests), but it is significant that it is EdF that has the reputation in the French political system of constituting a 'state within a state'. EdF however could not have risen to such power if its ambitions had not coincided closely with those at the highest levels in the political system. In this process, the Ministry of Industry was important less for its independent action than for its pivotal position. It was the place where the concerns of president, prime minister and the Council of Ministers about strategic matters such as security of supply and external dependence could encounter the attractive and well-organised programmes offered by EdF, the CEA and PEON. It is worth remarking here that the centralisation, social homogeneity (and close personal ties) within the upper reaches of the French political system made the process of reaching an agreed strategy rather less problematic than would have been likely in most other Western countries.

The problem was essentially to gear up a system which had been working at a level of about 1 GW per year of completed plant to one capable of putting 6 GW of plant a year into commission. A key early requirement was to build up industrial capability particularly in the Framatome group channelling effort away from reactor development, such as work on larger reactors and designing new steam generators, towards increasing productive capability. (See Masters (1979, pp. 54–61) for further details on the restructuring and the investment in productive capability.)

All major components with the exception of the turbine generator (supplied by Alsthom Atlantique and CEM) were to be supplied either directly by Framatome, for example pressure vessels, steam generators and pressurisers, or by other companies from within its then parent Empain–Schneider Group, for example primary pumps supplied by Jeumont Schneider, primary piping supplied by Spie–Batignolles. The Framatome production facilities were concentrated at two new factories capable of supplying components for about eight reactors a year. These two factories, one at Le Creusot supplying pressure vessels, and one at Chalon-sur-Saône supplying steam generators, were committed in 1971 and when, in 1973/74, the French government and EdF began to discuss a programme of about 12 plants, Framatome increased investment particularly in the Chalon facility. By the end of 1975 both factories were in production.

Considerable investment was also required in turbine generator manufacture facilities. These were initially split between Alsthom, which won the orders for the CP 900–1 tranche, and CEM which won the orders for the CP 900–2 tranche. However, in 1977 Alsthom bought out CEM and since then all orders have been placed with Alsthom.

Anticipating a lucrative world market in reactor sales and foreseeing the home demand for reactors could not be sustained at this high level, the French nuclear industry also sought to break into export markets. This put them into direct competition with a number of more experienced vendors including suppliers from the USA, FR Germany, Canada and Sweden bidding for a small number of units. Nevertheless the French industry did have a number of advantages over its competitors including co-ordinated strong national support at government level and a more comprehensive array of capabilities than any other supplier could offer.

The national support, in which the Ministry of Industry was the key actor, was important in bringing together French industry, providing finance and in smoothing any problems arising from proliferation considerations.

There were two aspects to the comprehensive supply of capabilities. For each order it was seen as important that France offer full plant supply (through Framatome and Alsthom), architect-engineering (through offshoots of EdF and the CEA), and operator training. In addition, France set out to offer full industrial fuel cycle facilities, from enrichment and fuel fabrication to waste disposal and reprocessing. In the longer term, France aimed to develop a world technological lead in FBR technology so that as uranium supplies diminished and FBR technology became more attractive, France could, on the basis of her reprocessing and FBR technology, take a large share of an expanding market.

However, the late 1970s was a period when few export orders for nuclear plant were placed and a controversial order for two reactors for South Africa was the sole order to be placed during President Giscard's period of office (see Table 8.6), although groundwork was laid in a number of countries. Whilst in technical terms, this expansion of capability proceeded far more smoothly than might have been expected, public opposition intensified and at some points appeared to threaten the overall strategy. (For a more detailed account of public opposition in France see Nelkin and Pollak (1981).)

The 1970 programme created the roots of the anti-nuclear movement in France and demonstrations and occupations took place at the Bugey and Fessenheim sites in the early 1970s. However, it was only after the major expansion, announced in 1975, when a large number of sites was named and plans became more tangible that the movement began to flourish. One of the special factors that gave the movement extra impetus and specific form

Table 8.6. *The French nuclear exports*[a]

Country	Name	MW (gross)	Year of order	Year of commissioning	Turbine supplier[b]
Spain	Vandellos 1	596	1966	1972	Alsthom/JS
Belgium	Tihange 1	920	1968	1974	Alsthom/ACEC
Belgium	Doel 3	936	1974	1982	Alsthom/ACEC
Belgium	Tihange 2	941	1974	1983	Alsthom/ACEC
Iran[c]	Karun 1 & 2	2 × 900	1974	—	Alsthom
South Africa	Koeberg 1 & 2	2 × 965	1976	1984, 86	Alsthom
South Korea	Uljin 1 & 2	2 × 950	1982	1989	Alsthom
China	Daya Bay	2 × 936	1986	1992	GEC

[a]With the exception of Vandellos 1, all units are PWRs supplied by Framatome. For the Belgian units, Framatome collaborated with ACECO (Belgium).
[b]Turbine suppliers:
    JS – Jeumont Schneider
    GEC – General Electric Company (UK)
[c]The Iranian orders were cancelled in 1979 with a significant proportion of construction complete.
Source: SPRU nuclear power plant data base.

in France was the difficulty of influencing decision-making partly due to the centralisation of government. This had a number of manifestations:
– when the decision to increase the size of the nuclear programme was taken in 1974, there was no parliamentary debate
– procedures for local consultations were transparently a charade. Work on site had frequently started prior to the public inquiry and the information provided by EdF was summary and often incomplete

Opposition was strongest at rural sites where agricultural land was expropriated, although this was often balanced by support from blue-collar workers and local traders who would benefit from increased job opportunities and trade.

Over the period 1975–77 demonstrations grew larger and more violent, culminating in the death of a demonstrator at the site of the Super Phénix FBR and the explosion of bombs at the Fessenheim site. It was perhaps these events coupled with the failure of the demonstrations to effectively influence policy and the failure to win legal battles that took the steam out of the more violent opposition. Opposition was increasingly channelled through political parties, trade unions and pressure groups.

The Socialist Party and its closely allied trade union, the CFDT, whilst never overtly anti-nuclear as a matter of principle, expressed reservations on a variety of grounds. They were particularly concerned about the potential over-capacity in electricity supply and the imbalance in energy policy. They felt that the resources devoted to nuclear power were adversely affecting other, sometimes more attractive, options such as conservation, coal and renewables. In addition CFDT campaigned vigorously for improved safety measures for workers at nuclear installations, especially at the troublesome La Hague reprocessing plant, publishing one of the most effective critiques of French nuclear policy in 1980 (Syndicat CFDT).

By contrast, its potential political partner in government, the Communist Party and its allied trade union, the CGT (Confédération Générale du Travail), were strong supporters of the nuclear programme. It was divisions such as this that may have contributed to the denial to parties of the left of electoral success in the 1979 parliamentary elections.

The ecology parties, particularly les Amis du Terre (Friends of the Earth) also gained some notable successes in by-elections although, perhaps predictably, they were unable to carry this success through to the general elections. Support for the ecologists came from a large group of academic scientists, GSIEN (Groupement Scientifiques pour l'Information sur l'Energie Nucleaire), which published a regular journal, *La Gazette Nucleaire*, on technical and legal aspects of nuclear power.

Whilst opposition through these groups and the 'cause celebre' of the proposed PWR site at Plogoff in Brittany continued to maintain nuclear power as a significant public issue, there was no visible impact on the programme or its method of implementation until the 1981 presidential election.

The 1981 general election was inevitably a turning point with a number of issues requiring action (regardless of which candidate had been successful).

Over-capacity in electricity supply. Whereas in countries such as FR Germany, Canada and the USA the large expansion in nuclear capacity envisaged in the wake of the 1974 oil crisis had been constantly revised downwards from 1976 onwards, in France the plans had been adhered to remarkably closely up to 1981. By then, it was apparent that, with six reactors a year being completed and little slackening in the pace of ordering, demand was highly unlikely to grow fast enough to continue to ensure reasonable utilisation of plant.

The indebtedness of EdF. The strains of financing the nuclear programme whilst still maintaining or even reducing the real price of electricity – to raise prices would have depressed demand and damaged the credibility of nuclear power as a cheap energy source – was causing EdF severe financial difficulties. Debts were mounting, many of which were incurred on the international market, and large operating losses continued.

Public participation. Although the violent demonstrations of the 1970s had largely subsided, it was evident that there was public dissatisfaction with the autocratic nature of decision-making which allowed little or no outside influence on the official plans.

Of the major party candidates at the election, only the Socialist candidate, Francois Mitterand, stood for major changes in the nuclear programme with the other three (two right-of-centre and one Communist) supporting a continuation of existing policy. The Socialist position was somewhat equivocal but appeared to encompass a number of main points:

– a moratorium on nuclear orders pending a national debate followed by a referendum
– a re-evaluation of the fast breeder programme
– an increase in state ownership in the heavy electrical industry
– an expansion of indigenous coal production and renewables, and more emphasis on conservation.

This latter objective of 'better times for all' except nuclear power (existing reactors would nevertheless be completed greatly increasing the nuclear contribution) was only possible by assuming economic growth rates of about 5%.

It had been expected that no party would win a decisive victory but, in the event, the Socialists did just that leaving the nuclear industry in a state of nervous uncertainty.

THE MITTERAND ERA (1981 ONWARDS)

Despite the apparent mandate to act strongly to bring the nuclear industry under democratic control and into better balance with the rest of energy and economic policy, President Mitterand's actions have fallen far short of the hopes of some of his supporters.

The immediate reforms

The national debate shrank to a two-day parliamentary debate resulting in a policy which differed little from that which probably would have been followed by the previous government. Reactor orders for 1982–83 were reduced from nine units to six units, and reprocessing and the FBR programme were continued essentially unaltered. Promises were made about more effective local participation in decision-making about siting.

Public ownership. The promise to increase public ownership in Framatome was fulfilled although not due to any concerted policy action by the government. Framatome's parent group, Creusot–Loire was already in financial difficulties and in the autumn of 1981, the CEA increased its holding from 30% to 34% and EdF's financial problems required Framatome to allow EdF easier financial terms.

The government/industry interface. At the government/industry interface a number of steps were taken to reduce the centralisation of decision-making especially reducing the role of the Ministry of Industry. Responsibility for the CEA was shifted to the Ministry of Research and Technology and the Energy Ministry and Planning Ministry were to take a more active and visible role in decision-making.

Planning procedures. One of the firmest commitments of the new government was to allow much greater local participation in site-selection. As evidence of good faith in this respect, plans to proceed at Plogoff, scene of considerable opposition, were abandoned. Work at five sites, Golfech, Cattenom, Chooz, Le Pellerin, and Civaux was suspended during the summer of 1981 pending the 'nuclear debate' although the three latter sites were in fact not actively being developed at the time. Perhaps more significantly, work was not interrupted at Nogent and Penly because of local opposition to any delay at these sites.

The municipal councils within a 5 km radius of the proposed sites were asked to give their opinions on the development of the sites although they were only given a matter of weeks to give their verdict. If they approved, the development work was allowed to proceed but if they did not, the regional authorities were consulted, with final authority residing at the National Assembly. Three out of five of the sites were approved at the local level whilst the regional councils gave approval for the other two sites. Ironically this placed the government in the situation of having more sites than required orders and eventually Le Pellerin, which for a number of reasons was not considered a good site, was abandoned.

Subsequently the regional authorities have played a stronger role in this process and, particularly in regions where the decline in traditional industries has led to serious unemployment, they have often actively lobbied for the siting of a unit in their region as a means of regional development. An example of this is the Civaux site in Poitou-Charentes.

Plant overcapacity. This problem was tackled immediately by reducing the ordering programme, although allowing more orders than seemed to be justified on capacity-need grounds. In addition some means of stimulating electricity demand both for French use and as exports were seen as desirable. The October debate reduced the volume of orders for 1982 and 1983 by one-third, based on a predicted growth rate of 5% per annum in electricity demand. It rapidly became clear that even this reduced rate was too high. A group within the National Planning Commission, the Energy Advisory Working Group chaired by Noël Josephe, was set up in 1982 to examine demand and the appropriate ordering rate for nuclear stations.

Inevitably with a group drawn from a wide section of interests, both internal and external dissent was considerable. On the one side were the Socialists and the CFDT whilst all other interests were lined up against them. The Energy Ministry, EdF and the CEA argued that the Josephe group underestimated demand growth – the report used a forecast of 350–370 TWh in 1990; the industrial interests, primarily Framatome and Alsthom argued that unit costs would be increased by reduced ordering and that expertise would be lost – they claimed that the minimum ordering rate was three units per year; the Communists and the CGT argued that the forecasts of GDP growth were too pessimistic; the highest scenario foresaw GDP growth of 2.2% per annum until 1990 and 4.6% from 1990 to 2000.

An interim report, published in October 1982, hinted that ordering would need to be severely reduced and the report submitted to the government in May suggested that simply to meet demand, no new orders would be needed until 1991. However, to maintain industrial capability the report suggested an ordering rate of one unit a year. This report brought a storm of protest both from within and outside the working group and the final report issued in July 1983 brought forward the date at which reactors would be required on demand grounds to 1987 although maintaining the recommended programme at one unit per year.

The government compromised and reduced orders for 1983 from the previously agreed level of three to two units, with two units to be ordered in 1984 and one or two in 1985. Ordering plans for the rest of the decade were not specified.

Subsequent to these immediate reforms, the French nuclear industry has been undergoing a number of adjustments and consolidations, notably:
- the completion and bringing on-line of the orders placed in the late 1970s (analysed in detail in a following section)
- technological developments by the nuclear industry asserting its independence from its licensors
- the struggle for positions of influence within Framatome
- measures to stimulate electricity demand growth
- an annual bargaining session between the government, Framatome and EdF to determine how many units should be ordered

These adjustments took place against a world background of deep recession which brought economic growth in almost all countries to a standstill (severely affecting electricity demand). The already slender flow of potential export orders for nuclear power stations practically dried up and fossil-fuel energy prices fell sharply in real terms.

Technology developments. Framatome terminated its licence with Westinghouse in March 1981, about a year before it was due to expire, and signed a new agreement on exchange of research and information to run until 1992. This new arrangement, in which Creusot–Loire took over Westinghouse's remaining 15% share, had three main advantages over the previous licence:
– licensing fees to Westinghouse which previously represented about 5% of Framatome's costs were replaced by a flat rate system of fees
– Framatome was free to sell its product worldwide under its own name without reference to Westinghouse or the US Government
– Framatome was free to make whatever changes it felt appropriate to the design

This formal break with Westinghouse was in fact the culmination of a long process whereby the design of each tranche of orders incorporated a higher French design content. In 1978, design work was started on the N4 series (see Table 8.5) which was to be the first PWRs which could be described as wholly French in design. Two factors underlay this decision:
– to assert Framatome's independence from its licensor
– to embody additional safety features then being requested by the French safety authorities (in 1979 the lessons from the TMI accident were added to these requirements)

This was the first tranche of French reactors which benefited from the earliest design stages from the actual operation of the first Framatome units. By 1983, this design work was complete and, in 1984, the first order was placed. Concurrently with this design work at Framatome, the turbine generator manufacturer, Alsthom, was also freeing itself from its licence agreements with Brown Boveri and for the N4 series, it proposed a fully French design of turbine generator, capable of utilising the increased output of the N4 reactor, equivalent to about 1500 MW.

In addition to these design developments, the operation of the first tranches of reactors led, by necessity, to the development in both EdF and Framatome of a wide range of skills in the reactor servicing area particularly in systems which have proved problematic such as the steam generators and control rod drives. The failure of demand to match the rate of ordering has led to the need to develop methods for reactors to 'load-follow' (i.e. allow fluctuations in output at short notice). These capabilities in servicing have opened up markets to French industry worldwide, ironically especially in the USA in competition with its former mentor, Westinghouse.

The power struggle for Framatome

The increasing financial difficulties of Creusot–Loire (little related to the performance of Framatome) led to its bankruptcy in 1984 leaving the CEA,

by default, as the sole stockholder of Framatome. A number of groups had been vying for greater control and influence over Framatome and the government was faced with a number of bids, all wishing to take up some or all of Creusot–Loire's holding. The main contestants were:
- the CEA, anxious to maintain its pervasive influence on all aspects of nuclear power
- EdF, interested in being able to increase its influence over the prices it pays
- CGE, as a complement to another of its subsidiaries, Alsthom which supplies the turbine generators. This would give CGE a comprehensive capability in nuclear power station supply
- Alsthom, with motivation as above. This proposal was most unpalatable to Framatome as it would have placed them as subservient to what it regarded as its junior partner
- Dumez, a construction group with strong Middle East connections
- Bouygues, France's largest construction group which like Dumez was already involved in the French programme
- Framatome management, keen to increase their say in Framatome's future direction

The government's decision, announced in August 1985, was to distribute the holding amongst five groups; the largest share, 40% went to CGE, the CEA's stake was reduced to 35% with the rest being distributed between Dumez (12%), EdF (10%) and Framatome management (3%).

Measures to stimulate electricity demand growth*

A massive effort to stimulate electricity demand growth was undertaken both by EdF and the government (through its energy conservation agency AFME). In the early 1980s, what demand growth there was, was entirely accounted for by increased electricity consumption at the uranium enrichment plant at Tricastin but in the last year or two demand has grown in all sectors (variability in weather conditions from year to year makes it difficult to calculate the true growth rate over a short period). Much of this growth has been in the crude heat (including space-heating) market. A wide range of tariffs dependent on time of day and time of year has made a number of applications apparently economic which would not have seemed possible a few years ago. For example it is now economic for steam boilers in industry to be electricity-fired during the summer months.

The stimulation of the space-heating market (electricity now accounts for nearly two-thirds of the space-heating systems in new dwellings) has

---

* See for example Fouquet (1985) and Jestin-Fleury (1985).

brought with it a peak load problem. In the winter of 1984/85, despite having a considerable surplus of nuclear generated electricity for much of the year. EdF had considerable difficulty in meeting its peak load. Whilst overall electricity demand grew by 7%, peak demand grew by 25% in 1985.

As mentioned above, the most successful area of demand growth stimulation has been for exports. By 1985 nearly 10% of French electricity production was exported, mainly to Italy and exports to the UK and Spain are growing quickly. The extent of this growth has been such that transmission lines have become heavily loaded and their capacity may be a constraint on further growth.

Rate of ordering

The rate of ordering of new plant has proved rather more difficult to control than might have been expected. Both the government and EdF have been conscious of the long-term need for a strong reactor supply industry if only for their reactor servicing needs, and have been reluctant to cut off the flow of orders entirely although, arguably, none was needed. The PWRs ordered by the end of 1981 alone, if all operated at base load, would produce more electricity than was consumed in France in 1985. In fact ordering has now settled down at one reactor per year and is unlikely to rise above this level in the near future.

THE STANDARDISATION PROGRAMME

The French programme of standardised reactors is unprecedented in its scale and degree of standardisation and it is therefore worth examining in detail how it was carried out, in particular examining what design changes there have been, how they were incorporated, how the programme was financed, and its impact on EdF's finances.

The need for standardisation

Given the size of the French programme, EdF argued that a high degree of standardisation was not only desirable but essential. This was said to be for a number of reasons:
- the work-load on regulators would be too great unless there was a high degree of commonality between plants
- design resource constraints meant that development work on only a limited number of plant systems could be carried out at any one time
- given the pace of ordering, it was inevitable that any design change would be incorporated in up to 30 units before any operating experience with the

new system had been gained. Thus, to minimise the risk of a large programme incorporating a serious design fault, it was necessary to incorporate design changes in a conservative and incremental way. This view of course depended on a high degree of confidence that the Westinghouse design was itself free of fundamental defects.
– over such a large programme, with such major implications for the French economy as a whole, it was essential that costs be kept as low as possible

Standardisation ought to achieve this last objective in two main ways. First, components could be manufactured in long and predictable runs, achieving low unit costs. Second, since units are essentially identical and built several to a site, learning in construction scheduling and the building up of a skilled and experienced labour force should allow short construction times, which also generally reduce costs. Further advantages in the use of identical components include the ability to change construction schedules by the 'lending' or 'borrowing' of components between units and the minimisation of the inventory of spare parts that needs to be carried.

In addition to these factors, the fact that there was only one customer, EdF, meant that, unlike the USA and FR Germany where plant tends to be tailored for each customer's needs, there was no pressure to customise designs.

*The extent of standardisation**

Whilst, as is argued above, standardisation was to some degree not so much an option as a necessity, EdF has been rigorous in its pursuit of standardisation, to the extent that duplicates of the engineering drawings are used in the construction of a whole tranche of stations. Of course some features of the design must vary, for example the cooling system depends on the siting of the plant (sea, estuary, river or cooling tower are the options) as do the foundations. In almost all cases monopoly suppliers of components have been used and in some cases these suppliers have been specially created. Whilst this decision meant that the anticipated advantage of competitive pressures amongst suppliers was foregone, it was expected that this would be more than counterbalanced by the maximisation of scale benefits and the capacity of EdF's technical and commercial skills to keep prices down. One exception to this strategy of using monopoly suppliers concerned the supply of the turbine generators. In the settlement of 1975, this was split between CEM and Alsthom–Atlantique. Alsthom was awarded the contract for the CP 900–1 tranche whilst CEM was awarded the contract for the CP 900–2

* For further details, see Tanguy (1985) and Bell *et al.* (1985).

tranche. This anomaly was removed in 1976 when CEM's works at Le Bourget were given to Alsthom–Atlantique in exchange for a 6% shareholding in Alsthom–Atlantique and an agreement that Alsthom would supply the 1300 MW turbine generators under licence to BBC, the parent company of CEM.

Despite this strong commitment to standardisation, design changes have occurred, particularly between tranches. This has been for two main reasons:
– as new operating data and research studies became available, regulatory requirements changed
– Framatome felt that significant improvements could be made to the details of the Westinghouse design

In some respects the first tranche, Programmes '1970', was prototypical, embodying a number of design changes within the tranche and ordered over a much longer period than the later tranches. The first two units (Fessenheim) followed the earlier export to Belgium (Tihange 1) in using the Westinghouse Beaver Valley (USA) design as the reference. For the next orders (Bugey 2–5), Framatome took the North Anna (USA) design as reference. Significant detailed modifications were made to the core, the control system, the pressure vessel and the fuel configuration, but the steam generator design was the same as that used at the US Trojan plant (type 51A).

The Bugey design formed the basis of the largest tranche of plant, the 16 units of CP 900–1, but included minor 'tuning' changes to the fuel and the adoption of a later Westinghouse design of steam generator (type 51M).

For the third tranche (CP 900–2), the vendor and design of the turbine generator were changed although, as was discussed earlier, this was not on the basis of any known technical advantages. In addition the feed water plant was changed and the turbine hall was placed radially rather than tangentially relative to the nuclear island. This arrangement, which has been used for all subsequent orders, was designed to minimise the risk of damage to the containment from 'missiles' which might be produced if a turbine were to break up.

At the same time as the CP 900–2 tranche was being ordered, the CP 1300–1 tranche was being developed. Design work on the 'P4' series of 1300 MW reactors had been started in 1973 but had been put on one side to speed progress with the earlier tranches. In 1975–76 work was resumed and the first order (of eight in this tranche) was placed in 1977. The primary motivation for increasing unit size was to improve the economics by obtaining economies of scale. As previously, an American unit's design (South Texas) was used as a reference but the French design departed from

the reference rather more than previously. The main changes from the 900 MW design were:
- four rather than three coolant loops with a commensurately larger core
- an increase in the power output of each loop from 928 MW(th) to 954 MW(th)
- rationalisation of layout
- incorporation of new safety features

The main safety features to be incorporated were greater redundancy in safety systems. Whereas the 900 MW units were built in pairs and shared many auxiliary circuits, the 1300 MW units are being built singly with completely autonomous circuitry. The containment was also changed from a pre-stressed concrete containment with steel liners to a double-walled concrete containment. The control room incorporated new micro-electronic technology and attempted to learn from the lessons of TMI.

The increased safety requirements inevitably reduced the scope for cost reductions and meant that the effect of the rationalisations was masked. For the next tranche of orders, which to date numbers 12 units, the technical characteristics remained unchanged but the layout was further modified to reduce costs. The design was designated P'4 and was the first design that could benefit from operating experience at a modern French PWR (Fessenheim 1 had been completed in 1977).

Not all the concerns of the safety authorities could be included in this design and this factor along with a desire to assert the independence of Framatome from Westinghouse led Framatome to commence work on a more radically different design designated N4. This allowed them to incorporate their own experience of design, building and operation more fully than before and also the opportunity to embody lessons from TMI early in the design process.

The most obvious change was the output increase from about 1300 MW(e) to about 1450 MW(e). This was made possible by raising the power of the core and increasing the output of each coolant loop from 954 MW(th) to 1068 MW(th). The steam generator design was changed to incorporate an economiser (increasing efficiency) and the tube material was changed from Inconel-600 to Inconel-690 to improve corrosion resistance and reduce occupational radiation exposure. Despite the higher power output the new steam generators were lighter and smaller than the previous design.

In consequence of these design changes the reactor vessel was enlarged, the primary pumps' efficiency and capacity improved and a new turbine generator design ordered. This new turbine, the Arabelle, will be the largest in the world. The reactor has also been designed to allow easy load-following capability, and to be able to operate for 17 months continuously between refuellings.

Finance

Finance for the programme has been a considerable burden on EdF and on the French economy as a whole and by 1982 EdF debts amounted to more than 150 billion francs, about a third of which was in foreign currencies. (*Nuclear Engineering International*, 1982a, p. 4.)

The reduction in the pace of ordering has meant that the problem has not become more serious, although the fall in the French franc and the policy of reducing electricity tariffs have not helped finances. By 1985, EdF's indebtedness had reached 213 billion francs, equivalent to more than 18 months' revenue. (For further details on EdF's finances see *Nucleonics Week*, 1986a, p. 7.)

During the late 1970s and early 1980s, EdF made heavy trading losses due to a number of factors other than the burden of heavy interest payments. Over the whole of this period tariffs were kept low partly as a counter-inflation measure and partly to reflect perceptions that nuclear power would lead to lower long-run marginal costs. In addition, in the period from 1981 to the end of 1983, the franc declined by nearly 50% against the dollar increasing the debt costs (often financed in dollars).

By 1985, the franc had recovered somewhat and increased revenues from home demand, which grew by 7%, and from electricity exports, which earned a surplus over import costs for the first time ever, meant that EdF made a trading profit for the first time in eight years.

ECONOMIC PERFORMANCE*

Operating performance

The French stock of reactors is very much more homogeneous than that of other countries in terms of design, vendor and electric utilities. Thus variations in performance between different tranches and through time can be much more unequivocally attributed to technological and maturity effects than for example in the USA where the large number of utilities, vendors and architect-engineers makes such attributions very risky.

The major feature distinguishing the way in which French PWRs are operated from any other set of reactors in the world is the degree of 'load-following' they are called on to undertake. Capacity has out-stripped demand to the extent that the PWRs must be operated in load-following mode for up to eight months of the year. This appears to have reduced capacity factors by about five percentage points below capability.

---

* Since they now represent obsolete technology, the French gas/graphite reactors are not included in this analysis, nor is the early Chooz PWR which is not representative of current designs.

Table 8.7. *Performance of mature French PWRs by time period*

	Programmes '1970'		CP 900–1		CP 900–2	
	Capacity factor	No. of reactor years	Capacity factor	No. of reactor years	Capacity factor	No. of reactor years
1980–82	61.7	8	–	–	–	–
1983–85	73.0	18	76.9	31	–	–
1986	72.3	6	76.7	16	80.2	2

Source: SPRU nuclear power plant data base.

The extent of load-following and the reactors affected are discussed in detail in a later section but this load-following has two important economic consequences. Firstly, this reduction in capacity factors means that the capital charges must be spread over fewer units than they might otherwise have been. Secondly, replacement power costs for much of the year may be relatively low or even zero. If the marginal source of power on the grid is to increase the output level of PWRs, then the replacement power costs are zero. In the spring and autumn when the PWRs are only required to reduce output for a few hours a day then the replacement power costs are likely to be those associated with operating modern efficient coal-fired plant. However, because of the peak load problem discussed earlier, replacement power costs on the coldest winter day could be very high, even those associated with importing power.

The operating performance of French PWRs is summarised in Table 8.7 and Table 8.8 and is shown in greater detail in the Appendix to this chapter. The main points to emerge from these analyses are:
– overall the operating performance of the full programme has been good
– the reactors' early performance is, in terms of capacity factors and shutdown frequency, somewhat below the settled-down level particularly in the second year when a major plant review and the first refuelling is scheduled. This suggests that there are significant installation faults and minor design errors which must be rectified
– the performance of each successive tranche appears to improve on the previous ones
– no significant derating has been necessary and operating losses are reasonably low
– shutdowns due to operating errors, training and licensing and regulatory limitations have been negligible. The relatively large amount of time lost for 'other' causes was the result of a variety of factors with no particular pattern

Table 8.8. *Performance of French PWRs by unit age*

Age	Capacity factor (no. of reactor years)				
	Programmes '1970'	CP 900–1	CP 900–2	Tranche 900	CP 1300–1
1	64.2 (6)	64.2 (16)	70.6 (8)	77.0 (2)	49.9 (2)
2	65.6 (6)	61.9 (16)	79.4 (6)	64.7 (1)	—
3 onwards	70.0 (32)	76.8 (47)	80.2 (2)	—	—

Source: SPRU nuclear power plant data base.

Three technical areas are worthy of more detailed examination; the identification and remedy of generic problems; steam generator performance; and load-following.

Generic problems

To date there have been four generic problems, that is problems that arose from a design or manufacturing fault which have affected an entire tranche or more of plant. These were pressure vessel cracking, control rod drive mechanism failure, turbine generator vibration and in-core instrumentation thimble faults.

*Pressure vessel cracking.** This problem arose during 1978, and, since it was detected prior to the start-up of the tranche of reactors it affected, it is not reflected in the operating records of reactors. The problem affected about 12 reactors and delayed the start-up of a number of these units.

As a result of an error in manufacturing procedure in the cladding process – the process in which a thin layer of corrosion-resistant material is added to the pressure vessel – a number of cracks were created in the pressure vessels near the nozzles. This is a potentially serious problem because if a crack in the pressure vessel exceeds a given size (this varies according to the position of the crack) the pressure vessel will rupture and expose the core without prior warning leakages. Three aspects of this problem are significant:

– although Framatome and EdF first became aware of the problem late in 1978, it did not become public knowledge until September 1979 when a French trade union, the Confédération Francais du Travail (CFDT) discovered and announced the cracks' existence

---

* For further details of the nature, causes and remedies for the problem see Wilke (1980), pp. 27–9.

- the start-up of three units, the first of the CP 900–1 tranche was delayed
  by three to six months as a result of pressure brought to bear by the CFDT
- Westinghouse were apparently not aware of the divergence from their
  procedures (see Select Committee on Energy, 1981, p. 453).

After a delay of about six months during which a number of tests and studies
were carried out, fuel loading was allowed at the affected reactors and it is
now expected that if any remedial action is required to repair the cracks it
will not be until fairly late in the life of the reactors.

The cost of this cracking was mostly borne by EdF in the form of
replacement power costs. At the time, EdF had relatively little spare plant
and it was hoped that the three units would have made some contribution
to meeting the 1979–80 winter peak and reduced the utilisation of EdF's
highest cost plant. The costs therefore to EdF were by no means negligible.

Control rod drive mechanism. * This fault was the most serious problem to occur
on operating plant and affected all of the Programmes '1970' and CP 900–1
tranches. The control rods are driven into the core in order to shut down the
reactor and are an important part of the control system for any reactor. The
problem first became apparent in January 1982 at Gravelines B1 when one
of the pins that holds the control rod guide tubes broke off and entered the
primary circuit. Subsequently three other units suffered a similar problem,
with metal fragments ending up in the steam generators.

In fact EdF and Framatome had been alerted to the potential problem in
1978 when a similar incident occurred at a Japanese plant, Mihama 3.
Following this incident, research was carried out in France into design and
materials to avert the problem. What is now considered the optimal
arrangement is embodied in Le Blayais 2, half way into the CP 900–2
tranche. For three of the oldest reactors affected, EdF replaced the entire
guide tubes including the pins, whilst a technique was developed to simply
replace the defective pins. This latter method would drastically shorten the
repair time from about a year to a few weeks as well as avoiding a disposal
problem for the scrapped tubes.

In the event the operation did not proceed as smoothly as planned. The
pin changing machine was designed to fit the later units of CP 900–1 and
did not fit the early units of CP 900–1 which differed slightly in design. A
somewhat less elegant compromise was then adopted. This involved repairing the tubes and then refitting them to another unit on which the repair was
carried out later.

---

* For further details, see *Nuclear Engineering International* (1982b), p. 2, and *Nucleonics Week*
(1983), p. 7.

Turbine generators. The turbine generators built under licence from Brown Boveri for the CP 900–1 and CP 1300–1 tranches have encountered a number of problems during the early phase of their operation. The main problem has been vibration between the high-pressure and low-pressure sections of the turbine. There has also been vibration in the 1300 MW units in the shaft coupling between the turbine and the generator. The main effect of these problems has been to lengthen the period between first criticality and commercial operation particularly in the early units of each tranche for which the solutions to these problems had to be derived.

In-core instrumentation thimbles. These components, which are sited in the core were found to be vibrating unacceptably in the first units of the CP 1300–1 tranche. Although this problem was said to have no safety implications and was simple to rectify on the reactors still under construction it did, along with turbine generator problems delay the commercial operation of the first two Paluel units and the completion of the later units. In fact the Paluel units took about 18 months to move from first criticality to full commercial operation, a phase which normally takes about six months.

Steam generator performance

One of the major uncertainties that has faced the French nuclear programme has been the performance of the steam generators. From the earliest stages, it was known that the Westinghouse design had deficiencies and a French design, the Trepaud design, was developed and available at the time of the early orders. However, the conservative route of staying with the Westinghouse design was chosen, which, whilst not perfect, was a reasonably well known quantity with worldwide experience. Opting for an untested design which would be fitted to many reactors long before significant operational feedback was available, was thought to be too risky.

Being aware that the design was problematic, the French have developed contingency plans for replacing steam generators should it be necessary (any design of steam generator so far used could theoretically be fitted to any French PWR) (see Sort, 1985, pp. 45–47 for further details). More importantly there has been intensive effort researching methods of manufacture, materials and material treatment, water chemistry, methods of monitoring and detecting problems and methods of repairing faults. Occasionally mistakes have been made, for example manufacturing faults occurred to some of the reactors in the CP 900–1 tranche and stress corrosion cracking is proving a significant problem at some units (see *Nucleonics Week*, 1986b, p. 2). Nevertheless, Framatome and EdF, in order to deal with these problems, have developed techniques such as roto-peening and shot-peening which reduce the stress in the material arising from the manufacturing process and which contribute to stress corrosion problems. These techniques are proving highly marketable in the worldwide reactor service market.

For the future, the N4 series embodies steam generators which for the first time depart markedly from the Westinghouse design both in materials and in design. The performance of these steam generators will be a true test of whether the French can match the achievement of Kraftwerk Union's PWR where steam generator problems have been minimal.

Load-following

Since no other country in the world requires its reactors to load-follow to the extent that France does, Framatome has, by necessity, had to carry out pioneering work in this area (see *Nucleonics Week*, 1984a, pp. 3–4). The basis for load following is the use of 'gray' control rods that dampen power distribution disruption in the core during rapid changes in the power level. These control rods were fitted from new on all reactors entering service after September 1983 and the earlier reactors were backfitted with them during 1984 and 1985.* The N4 reactors have been designed specifically to be able to operate in load-following mode and should be able to accommodate more frequent changes in power level than the earlier units.

Construction times and capital costs

The nuclear industry internationally has made much of the rapid construction times and apparently low capital costs of French nuclear power. Whilst the available data allow for a reasonably full analysis of construction time experience, reliable information on the costs of French reactor construction is extremely limited.

It is possible to try to explain construction times and their variations in terms of a number of possible factors: Table 8.9 starts with the basic data, organised, as for other chapters, by year of construction start.† For 900 MW

---

* KWU disputes Framatome's claim that the load-following capability of its reactors is unique and claims that its own reactors are capable of similar performance but are not required to do it. (See *Nucleonics Week*, 1985, p. 9.)

† It should be noted that the French Government has changed its reporting of construction start dates to the IAEA. Between 1982 and 1985, the reported date of construction starts was changed for all reactors started before 1982, with only two exceptions (all completion dates are unaltered). These changes have no net effect on the overall average construction time, and we have used the most recent data here. However, one point to note is that the new reporting procedure reduces average construction time for the first tranche by almost six months compared to the earlier convention.

Table 8.9. *Construction times[a] of commercial French PWRs (months)*

Year of construction start	Unit	Construction time	Average construction time (no. of units)
1. 900 MW units			
1971	Fessenheim 1	76	
	Fessenheim 2	76	
1972	Bugey 2	75	75 (4)
1973	Bugey 3	71	
1974	Bugey 4	61	
	Bugey 5	67	
	Tricastin 1	73	69 (4)
	Tricastin 2	73	
1975	Tricastin 3	73	
	Tricastin 4	77	
	Gravelines B1	69	
	Gravelines B2	69	
	Gravelines B3	66	70 (9)
	Dampierre 1	67	
	Dampierre 2	70	
	Dampierre 3	68	
	Dampierre 4	71	
1976	Gravelines B4	66	
	St Laurent B1	88	80 (3)
	St Laurent B2	85	
1977	Le Blayais 1	59	
	Le Blayais 2	73	76 (4)
	Chinon B1	83	
	Chinon B2	89	
1978	Le Blayais 3	67	
	Le Blayais 4	66	69 (4)
	Cruas 1	67	
	Cruas 2	75	
1979	Gravelines C5	64	
	Gravelines C6	64	71 (4)
1980	Chinon B3	76[b]	
1981	Chinon B4	79[b]	
2. 1300 MW units			
1977	Paluel 1	100	97 (2)
1978	Paluel 2	93	
1979	Paluel 3	84	
	St Maurice/St Albans 1	88	
	St Maurice/St Albans 2	89[b]	87 (5)
	Flamanville 1	85[b]	
	Cattenom 1	90[b]	
1980	Paluel 4	74	
	Flamanville 2	79[b]	
	Belleville 1	88[b]	84 (5)
	Belleville 2	96[b]	
	Cattenom 2	82[b]	

units, perhaps the most immediate characteristic of Table 8.9 is that there seems to be no discernible trend in construction times over the decade covered. Construction times have varied, on average, within the narrow band of 70 to 80 months. Even more striking, however, is the fact that construction times have not risen over time. This, in the first place, is in direct contrast to the experience of the USA and FR Germany, where times have lengthened substantially during the 1970s. Second, it is all the more surprising given the very rapid expansion in the scale of the ordering programme. In the three years to 1973, only 4 reactors were started, while 13 were started in the next two years. Given that the programme was expanding so rapidly, it is remarkable that there were so few bottlenecks in either manufacturing capacity or, even more important, in site construction.

The 1300 MW units have on average taken substantially longer to build – 87 months for the 12 units complete (or nearly so) compared to 72 months for the 32 900 MW units. There is, however, some sign that times have reduced a little on the later 1300 MW units, though the number of units involved is small. The minimum construction time for 1300 MW units has so far been some 10–12 months longer than for 900 MW units, and EdF now argue that 1300 MW units will necessarily take longer to build, mainly because of the substantially greater civil works (Bacher, 1986, p. 41).

It is also of interest to look at the experience of construction times between, and within, the tranches in which French PWR construction has been organised. Table 8.10 shows that between tranches of 900 MW plant there has been remarkable consistency in average construction times at 71 or 72 months. It might be expected that given standardisation within tranches, learning would lead to shorter times at the end of tranche than at the beginning. However, Table 8.10 shows no consistent evidence of this – in two of the four 900 MW tranches the later units were built more quickly, while in the other two, later units took longer to construct. The very limited evidence in 1300 MW units is that there may well be some learning in later units of the one completed tranche.

The final factor that might be expected to exert a general influence on construction times is multiple units at a site. In the period covered by this analysis, all construction has taken place in combinations of either two or four units per site, and Table 8.11 shows construction times organised by

---

Notes to Table 8.9
[a] Defined as period from construction start to commercial operation (handover to utility).
[b] These are estimated times for incomplete reactors. All were close to commercial operation at the time of writing.
Source: IAEA (1985) *Nuclear power reactors in the world, April 1985 edition,* Vienna and *Nucleonics Week* (1985, 1986 and 1987), McGraw-Hill, various issues.

Table 8.10. *Construction times of different tranches of French commercial PWRs (months)*

Tranche	Average construction time	
1. 900 MW		
Programmes '1970' (1970–74)	71	
(6 units)		
First four units		76
Last two units		69
CP 900–1 (1974–77)	72	
(16 units)		
First eleven units		71
Last five units		74
CP 900–2 (1977–78)	72	
(8 units)		
First four units		76
Last four units		68
Tranches 900 (1979–82)	71	
(4 units)		
First two units		64
Last two units		77
2. 1300 MW		
CP 1300–1 (P4) (1977–80)	87	
(8 units)		
First five units		90
Last three units		81

Source: See Table 8.9.

unit number. Second units on the whole take slightly longer to build than first units, but this, according to EdF, is intentional. Site work is organised so that the first unit is on the critical path, and civil works actually begin earlier on the second unit than on the first (Bacher, 1986, p. 41). For the 900 MW units there does seem some evidence from Table 8.11 to suggest that learning effects on site mean that third and fourth units can generally be built more rapidly than first or second units. As yet there are no clear trends from 1300 MW units.

Having looked at general and average factors, it is necessary to look briefly at a few specific units where construction times have been somewhat longer than average, and at factors which have affected specific units. Among 900 MW units, only four – St Laurent B1 and B2, and Chinon B1 and B2 – have taken longer than 80 months to build. In the case of St Laurent units, construction to the point of first criticality was rapid: however the gaps between criticality and commercial operation were 31 and 27 months for units 1 and 2 respectively (against an average of around

Table 8.11. *Construction times of different units at same site (months)*

Site	Unit number 1	2	3	4
1. 900 MW units				
Fressenheim	76	76	—	—
Bugey	75	71	61	67
Tricastin	73	73	73	77
Gravelines B	69	69	66	66
Dampierre	67	70	68	71
St Laurent B	88	85	—	—
Le Blayais	59	73	67	66
Chinon B	83	89	76[a]	79[a]
Gravelines C	64	64	—	—
Cruas	67	75	—	—
Average times	72	75	69	71
2. 1300 MW units				
Paluel	100	93	84	74
St Maurice/St Albans	88	89[a]	—	—
Flamanville	85[a]	79[a]	—	—
Cattenom	90[a]	82[a]	—	—
Belleville	88[a]	96[a]	—	—
Average times	90	88	—	—

[a] These are estimated times for incomplete reactors. All were close to commercial operation at the time of writing.
Source: See Table 8.9.

six months). Chinon B1 and B2 also had longer than average gaps (16 and 11 months) between first criticality and commercial operation but also suffered earlier delays after the hydrotest stage (when construction is more or less complete). In all cases, modifications became necessary after construction was complete: this means that for 900 MW units physical construction times have been remarkably consistent.

As was made clear in the section on operating performance a number of technical problems – concerning pressure vessel cracking, control rod drive mechanisms and turbine generators – led to some delays in the construction of a few 900 MW reactors (e.g. Tricastin 1–3 because of pressure vessel cracking). However, these delays were minor and generally added only one to six months to construction times. In their absence it might have been the case that later units would on average have appeared to have been built marginally faster than earlier ones, though the effect would be extremely small.

For 1300 MW units there have been a number of technical problems which have somewhat delayed construction of the earlier units, especially on the first two units of Paluel. This led to periods of 19 to 16 months respectively between first criticality and commercial operation, but it remains the case that the physical construction process (up to first criticality) has consistently taken longer on 1300 MW than 900 MW units.

A final possible influence on construction times is deliberate stretchout of schedules because of financing difficulties and/or low load-growth. While EdF have been saying, since 1983, that such stretchouts are being contemplated for those reasons, the weight of adjustment has fallen overwhelmingly on the timing of new orders, which have become increasingly limited, and there is no firm evidence that stretchouts have significantly affected construction times of those units included in Table 8.11. Overall, then, construction times for 900 MW units have shown remarkable consistency and have been short (and with limited variation) by the standards of other countries covered in this book. There is little evidence of learning leading to reduction in construction times but this is at least in part explained by the rather short schedules that the French programme established from the beginning.

Available information on construction costs is disappointingly small. EdF has a monopoly on cost information and has, over time, released less useful data rather than more. The very global and aggregative form in which cost data are presented means that precisely what is included and excluded from EdF cost figures has never been clear, nor whether some of the subsidies offered by the French Government to EdF are reflected in EdF costings or not. This means that the absolute level of French construction costs must be treated with some care, if not suspicion, even for comparisons *within* France: and of course, international comparisons which rely on absolute numbers for costs are almost valueless.

This means that all that we can reasonably do is to check on the *changes* in French construction costs. A number of sources make it clear that real capital costs per kilowatt have risen by some 40% to 50% between the early 1970s and early 1980s. For instance, Bennett *et al.* (1982, p. 2) report a 43% increase in real costs between 1974 and 1981, while Crowley & Kaminski (1985, pp. 34–36) report a 54% increase over the 1974 to 1984 period.

These steady increases in real costs per kW are consistent with those found in the other countries studied in this book, somewhat higher than in the Canadian case but substantially lower than for the USA and FR Germany. EdF add a particular gloss to this experience in their explanation of the increases. They have constructed a cost curve for nuclear construction 'normalised to mid-1974 site and working conditions' (Bennett *et al.*, 1982, p. 22). This shows costs to be virtually constant over the 1974 to 1981 period, but comes dangerously close to (and possibly across the border into) tautology: if all major conditions had remained constant, it is hardly surprising if costs do not change either. In practice, site-specific conditions do seem to have led to a part of the cost increase (e.g. incorporation of cooling towers in inland sites) but tighter safety standards, including quality assurance requirements, also appear to have played a large part (Bennett *et al.*, 1982, p. 7). Interestingly, EdF expect further real capital cost increases in the future, and for the 15 years between 1977 commercial operation and 1992 operation, they expect 'fixed costs' (which are over-whelmingly construction-related) to increase by as much as 99% (Baumier & Bergougnoux, 1982). Again, safety and site-specific factors are cited as of major importance.

In the absence of useful data on individual stations, there are two further issues worthy of comment. First, there have been no economies of scale in 1300 MW units compared to 900 MW units, and indeed the (limited) 1300 MW experience so far is that 1300 MW units are marginally dearer on a per kW basis. This probably reflects the difficulties on early units, but EdF no longer expect 1300 MW units to be cheaper, though they clearly benefit from the potential saving in sites that larger unit sizes bring. Second, an interesting French/US study (*Nucleonics Week*, 1984, pp. 9–10) attempted to overcome some of the problem of international monetary comparisons by examining the comparative physical inputs into American and French PWRs. While the inputs of materials were closely comparable between the two countries they found an enormous difference in labour requirements. This difference is concentrated in the electrical/mechanical rather than civil area, and French PWRs use 50% less labour in electrical/mechanical work. The total of craft labour in French PWRs is only 35–65% of that needed in the USA, and most spectacular of all, a 4-unit French station needs, at 30m manhours, only marginally more total labour than a typical 1-unit US station. This strongly suggests substantially lower real capital costs in France than in the USA, and lends weight to the common belief that French capital costs have in real terms been relatively low, even if the information released about them makes it hard to establish this. In the domestic comparison between nuclear power and alternatives, the advant-ages to nuclear power have always been large on official figures (e.g. in 1983, even at a 10% discount rate, nuclear generation cost only two-thirds of coal-firing) (Nuclear Energy Agency, 1983, Table 11, p. 54).

However, this discussion of French capital costs has concentrated nar-rowly on the micro-economic aspects. While such a procedure may be justified for the other countries covered in this book, it is doubtful whether the nature of the commitment of the French state to nuclear power can be adequately analysed within a narrow micro-economic (pence/kWh) frame-work. A practical example of this problem is provided in the substantial involvement and assistance that French governments have provided in the reorganisation and investment in the capital goods industries (e.g. Framatome) which supply nuclear equipment to EdF. In economic terms, this means that low prices paid for nuclear equipment by EdF may reflect subsidies that are well 'up-stream' of the utility. Consequently, it is extremely difficult to establish whether French nuclear power has been as inexpensive as the official data show.

OVERALL EVALUATION

At least in energy policy terms, the objectives of the French nuclear programme can be readily summarised. These were to provide a cheap source of reliable indigenously produced energy which would substitute for imported energy. Specifically this would be done by substituting nuclear power for fossil-fuels in electricity generation and substituting electricity for the direct use of fossil-fuels.

Whilst it is possible to point to the reduction in petroleum use from 123 mtoe (67% of total energy) to 90 mtoe (48% of total energy) between 1973 and 1983, such measures do not really address the key questions about how successful the nuclear strategy has been so far. These would include:
- by how much and at what cost could oil consumption have been reduced without the nuclear expansion, for example by larger improvements in end-use efficiency, greater use of coal etc.
- to what extent has the nuclear programme distorted the French economy through pre-emption of investment capital and skilled manpower
- what competitive advantages have accrued to French industry through the availability of cheaper electricity than might otherwise have been available
- to what extent has France's vulnerability to external shocks been reduced and to what extent has France's vulnerability to internal shocks through the possibility of generic faults in the PWRs been increased

To answer these questions in any meaningful way would require a comprehensive analysis of energy demand, economic performance, and international oil politics far beyond the scope of this book. But they do suggest that while on narrowly economic grounds the French experience has been successful, a wider and more comprehensive evaluation might not automatically reach the same conclusion.

Turning to technical aspects, in many respects the French nuclear programme must be counted a remarkable technical achievement. In less than 15 years, the nuclear industry has been transformed from a fragmented group of companies into a highly co-ordinated effective industry capable of building reactors more quickly and cheaply than virtually anywhere in the world. It has achieved a high degree of technological independence. The CEA has similarly revitalised itself after its defeat over GCRs and is now firmly re-established in all aspects of the nuclear system, from its ownership of COGEMA, its role in regulation, its role in research and development, to its expanded interest in Framatome.

This achievement has only proved possible because of the coincidence of a number of necessary but individually not sufficient conditions:
- the use of a standardised design
- the existence of a highly centralised decision-making mechanism
- the strength and motivation of EdF and the CEA
- strong commitment and support from government

It is difficult to think of any other WOCA developed country with a political and economic system that could produce these conditions.

Paradoxically, despite this undoubted technical success, the overwhelming impression is still that more doubts and uncertainties remain than have been resolved. These centre on a number of points:
- the pros and cons of standardisation
- the ability of the nuclear supply industry to withstand the leaner times it is now facing
- the ability of the French nuclear power industry to innovate successfully
- the democratic accountability of the programme

Standardisation. As noted above standardisation has been an essential feature of the French programme. Its advantages in terms of cost control, construction scheduling, regulation, problem solving and repair scheduling have been rehearsed above. However, two outstanding points remain. First, over a period longer than about a year, standardisation to the ultimate degree is simply not feasible nor desirable, at this stage in technology development. Research and operating experience reveal problems with existing materials and methods, or better ways of achieving the same result, which it would be foolish to ignore. Very often, as in the case of pressure vessel fabrication or control rod drives, these changes in methods and/or materials have anticipated and reduced the scale of problems encountered later. However, there is always a risk, better illustrated in other countries, that the changes will prove counter-productive or, as in the case of the machine built to repair the control rod drives, frustrating. Second, and perhaps more importantly, the risk of standardised error remains real. Where, as in the case of the pressure vessel and the control rod drives, the problems are judged not to require immediate action, standardisation still has some positive advantages in terms of repair methods and scheduling to offset the cost of the multiple error. However, should the fault require immediate action, an unenviable choice would have to be faced.

Either a significant and perhaps crippling amount of generating capacity would have to be shut down, or the risk both physical and also to public acceptability would have to be carried, of continuing to operate the plant.

The resilience of the nuclear industry. Despite having the fullest order book of any vendor in the world, Framatome may be more vulnerable than its competitors in FR Germany or Japan. The decision to use single suppliers and to carry out manufacture of many of the major components within the company gave Framatome advantages in cost and control. However, now that the rate of ordering has declined, it is faced with a great deal of underused and expensive capacity and unpopular redundancy decisions. These decisions will be made harder to take because of the high level of state control and the national prestige that surrounds the nuclear industry. By contrast, in FR Germany, manufacture is subcontracted to large strong diversified privately owned companies which, in periods of low ordering can perhaps cross-subsidise and redeploy manpower. The vendor, KWU, itself part of a large strong company, has a relatively small amount of resources, mainly in design and software, which it must maintain in periods of low ordering.

Ironically, the relative absence of either design changes during construction, or regulatory hostility has meant that, unlike FR Germany where such factors have kept design resources busy late in the construction phase, in France, the design work is finished very early and any drop in ordering is rapidly felt.

It seems highly unlikely that exports could have anything other than a marginal effect on the situation. As is argued in Chapter 10, the total volume of nuclear exports placed with all vendors is likely to remain very low compared to the historic rate of French ordering, and, in the contests to date, France has not yet demonstrated that it has such a competitive advantage over the other vendors that it can expect to win more than a reasonable share of these orders.

One factor in its favour is that the scale of the programme and the closeness of the design to the Westinghouse original has meant that Framatome and EdF have built up a wide array of skills in the reactor servicing market which are already proving highly marketable worldwide. This should ensure a flow of income to Framatome and maintain meaningful tasks for its design teams, although of course doing little to maintain manufacturing capability.

Innovative capability. Much of the technical success of the French programme to date has been based on the strategy traditionally adopted by the Japanese of copying the Americans but doing it better. Whether the French can now go on and successfully incorporate their own innovations as the Germans have already proved themselves capable of doing will be tested by the operation of the N4 plants which are the first to incorporate a major French design content.

In the long term it is likely that no technology can survive unless it has wide popular support and this may not be forthcoming if it is felt that decision-making is autocratic and remote. Whilst the Mitterand Government has, on the face of it, improved this, the changes may be more cosmetic than real. The number of units to be built is still decided centrally, there is no legitimate forum for the views of objectors to be thoroughly tested and, in the current recession, with high unemployment, the local consultations are effectively very similar to the sort of incentives, such as the building of local amenities and the offer of low electricity tariffs to those near nuclear sites that characterised the previous administration's approach.

APPENDIX. DETAILED ANALYSIS OF THE OPERATING PERFORMANCE OF FRENCH PWRS

Table A8.1. *Causes of lost output in French PWRs according to age*

Age	No. of reactor years	Capacity factor	Shutdown losses	Derating losses	Operating losses	Adjusted operating losses
Programmes '1970'						
1	6	64.2	28.5	1.0	6.3	9.0
2	6	65.6	28.4	1.3	4.6	6.4
3 onwards	20	68.4	26.1	1.6	4.9	6.2
CP 900–1						
1	16	64.2	29.3	0.5	6.0	8.2
2	13	58.6	33.6	0.9	6.8	10.0
3 onwards	18	75.6	17.5	0.9	6.1	7.3
CP 900–2						
1	2	74.5	20.8	0.3	4.3	5.4

Source: SPRU nuclear power plant data base.

Table A8.2. *Causes of shutdown in French PWRs according to age*

Age	No. of reactor years	Shutdowns (hours per year of operation)																	Shutdown frequency (no. per year of operation)
		1	2	3	4	5	6	7	8	9	10	B	C	D	E	F	G	H	
Programmes '1970'																			
1	6	0	0	9	0	39	9	3	223	5	15	2	1976	343	33	9	0	190	11.4
2	6	0	0	5	0	133	26	14	55	6	0	0	1869	618	19	0	8	54	12.4
3 onwards	20	584	6	2	0	63	104	19	26	30	0	4	1311	335	9	0	13	88	8.4
CP900-1																			
1	16	0	0	54	6	100	7	120	183	80	27	10	1827	370	28	0	0	196	13.7
2	13	341	0	99	17	222	80	102	149	41	17	4	2184	92	45	0	19	348	13.3
3 onwards	18	0	1	24	5	58	95	11	118	7	5	0	1174	56	19	0	0	94	7.3
CP900-2																			
1	2	0	0	0	0	79	0	6	38	7	13	0	1678	0	0	0	0	90	8.8

Note: Causes of shutdown are:

1 = reactor and accessories
2 = fuel
3 = reactor control system and instrumentation
4 = nuclear auxiliary and emergency system
5 = main heat removal system
6 = steam generators
7 = feedwater, condenser and circulating water systems
8 = turbine generator
9 = electrical power and supply system
10 = miscellaneous equipment failure
B = operating error
C = annual refuelling and maintenance
D = other maintenance and repair
E = testing of plant systems/components
F = training and licensing
G = regulatory limitation
H = other

Source: SPRU nuclear power plant data base.

CHAPTER 9

Lessons from the case studies

Introduction

Arising directly from the case studies a number of general conclusions and observations can be drawn which can be broadly classified into those relating to technology development and those relating to institutional responsibilities and structure. Under the former category the following are considered:

- pre-commercial development including the use of prototypes and demonstration plant and the appropriate rate of development
- the necessary conditions for the efficient incorporation of learning
- the differing strategies for transferring in technology
- the experience of standardisation
- the extent and nature of scale economies
- the applicability of the standard investment appraisal techniques

Within the second category we draw on the experience of the four case study countries to determine what the key roles and responsibilities are. We look particularly at the utilities, vendors and safety regulators and examine how these roles and responsibilities might be most appropriately allocated under various economic, political and cultural backgrounds.

Technology development

In the framework for analysis set out in Chapter 2, three important processes were identified: pre-commercial development, the incorporation of learning and the transfer of technology. It is on these three processes and the importance of standardisation, scale economies and economic appraisal techniques that this section concentrates.

Pre-commercial development. This is a difficult phase to evaluate because, by its nature, it will throw up technological errors and blind alleys of technology development. The simple diagram (Fig. 2.1) showing the progression through basic research, proof of principle, prototype, demonstration unit to first commercial design rather neglects this 'noise' – competing products, confusion about the technological direction etc. – that is inherent in this stage and suggests a smoother progression than is feasible. The judgement on the success of this phase must therefore be based not so much on how well the pre-commercial plant operates but how thoroughly the product is investigated and debugged, and how well the lessons are learnt. Intuitively it might be expected that some of the requirements of a successful pre-commercial phase might include that:

– the pre-commercial work on each design be carried out by as small a number of companies as possible to ensure that the knowledge gained is concentrated and retained and can readily be drawn upon
– the results of each stage including design, building and operation are fully evaluated before the next stage is commenced
– successive increases in size do not represent too large a step upwards
– more than one design or design concept is developed to provide some element of competition
– there is significant user input in the development of the technology to ensure that it meets the users' needs

In practice the experience of the case study countries only partially bears out these hypotheses.

In the USA a number of competing designs were developed (although it has been argued that only the light water reactors had a realistic chance of success). (See for example Roth, 1982.) However, whilst the steps up in size were relatively conservative – a maximum of a factor of three – each successive increase in size was carried out long before its predecessor had entered service. For example, when General Electric won its first order for a plant of over 1000 MW, the largest GE BWR operating in the USA was of 200 MW and when Westinghouse won its first order for a plant of over 1000 MW, its largest PWR in the USA in operation was of 175 MW. This large step up in size was made worse by the fact that, whereas with units of up to 600 MW, designers and users could benefit from experience of fossil-fired units of this size, the orders for 1000 MW nuclear plant predated the first order for fossil-fired plant of this size. In addition, whilst there was some continuity in choice of architect-engineers in the early stage (Stone & Webster for PWRs and Bechtel for BWRs), when the first wave of commercial orders was placed a number of completely inexperienced architect-engineers won business. There was inevitably no continuity amongst utilities as few utilities were large enough to justify a flow of orders and those that were large enough were often slow to be convinced of the need for nuclear power. The utilities were effectively 'sold' nuclear power, first by the government and then by the vendors.

Perhaps the most serious error at this stage, however, was the very sharp transition from pre-commercial to fully commercial status that the Oyster Creek order brought about. This had three important consequences. First, nuclear power plant had to be priced to compete with very cheap coal and oil. With hindsight, this was simply not a feasible goal, but the pursuit of it led to a number of false economies. Second, the declaration of full commerciality led to an impression that nuclear power was fully mature with no major outstanding technical problems. On this basis a number of utilities placed orders for plant with an inadequate appreciation of just how stringent the requirements for success were. Third, the intense competition that was triggered between the vendors meant that their efforts were concentrated on fulfilling a large backlog of orders and containing or even reducing costs rather than ensuring the product was free of basic design faults and materials problems.

In Canada, the situation was different, with the impetus for technology development coming primarily from the utility, Ontario Hydro. The vendor, AECL, did do some development on reactor concepts other than CANDU, but Ontario Hydro concentrated its interest in CANDU. With Ontario Hydro supervising both the construction and architect-engineering the pre-commercial experience was concentrated and retained. Scaling-up was carried out in relatively small jumps although, as with the American experience, construction work was started on the next step up before there was any experience of operation of the previous step.

For France and FR Germany the situation is more difficult to evaluate because the technology that was ultimately chosen was imported and earlier indigenous designs were abandoned. In this situation an important factor is the amount and quality of information passing from the licensor. The process of technology transfer is discussed later but it is worth noting the existence of experience of some pre-commercial units in both countries before the main programmes were launched. Such experience could allow the buyer to critically assess the information coming from the licensor.

In France, the 305 MW Chooz plant (a Franco-Belgian collaboration) entered service two years before the first commercial PWR orders. Both Framatome and EdF participated in this project and gained valuable experience from it. Subsequently, the six orders of the first tranche were placed over an extended period of four years and this gave Framatome and EdF an opportunity to absorb the design fully and to learn from the limited operating experience elsewhere. These factors along with the very close relationship between the single utility and the vendor has meant that the experience gained was retained.

In FR Germany, PWR and BWR development followed rather different routes from each other. For the BWR, ironically in view of the mistakes made in its development, the progress followed a seemingly ideal path with steady progression in sizes and full operation of each stage before the next stage was started. A small prototype unit (VAK, 15 MW) was built and operated under the auspices of RWE, the largest utility. Later, two demonstration units (Lingen, 268 MW and Gundremmingen, 250 MW) were ordered and were completed at about the time the next step up was made (Wurgassen, 670 MW). These orders were spread amongst several utilities, but, in view of the very close co-operation that appears to exist between utilities, this should not have been a major drawback especially as the vendor (then AEG) carried out its own architect-engineering. However, this simple chronology of events does not show the mistakes and changes in direction that occurred which, with hindsight, should have led to a pause before the ordering of Wurgassen.

The German PWR technology transfer process was much more abbreviated with the first step being a 300 MW plant (Obrigheim). This unit had only just entered service when the first unit of over 1000 MW was ordered. Siemens, the vendor, has always stressed the importance of the 20 MW MZFR heavy water reactor which they claimed was sufficiently similar to a PWR in concept that they were able to learn a great deal of relevance from it.

In Canada, France and FR Germany, the transition from 'infant' industry to fully commercial industry was more gradual than in the USA. Government support, often indirect, was maintained over a transition period, and nuclear power was seen as a strategic commitment which did not immediately have to show an economic advantage over its competitors.

To summarise, it seems that the conservative scheme of proceeding slowly by small steps earlier, while intuitively attractive, is neither a sufficient nor a necessary condition. Where, as with Canada, the basic design does not embody significant weaknesses and no major changes of direction are undertaken, a certain amount of short-circuiting might be justifiable. However, if, as in the USA, important design issues remain unresolved or, as in FR Germany with the BWRs, a major change of direction is undertaken, a more cautious approach would seem wise. Slowing the pace of product change may prove difficult in the face of pressures from enthusiastic marketing and R&D departments to create or maintain a competitive advantage by embodying the latest research results in new designs.

The issue of the desirability of having competing designs is also not resolved. In France and Canada, there has not been more than one design simultaneously commercially available for users to compare. In FR Germany, the two commercial designs, BWRs and PWRs, have for more than 15 years been supplied by the same company, blunting competitive pres-

Design A   *Design B*

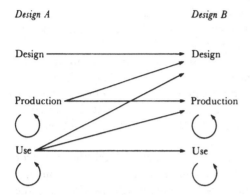

Figure 9.1. The incorporation of learning. The arrows represent the feedback of experience.

sures. In the USA, competition between vendors does not seem to have been a positive influence although it was perhaps the timing of commercialisation rather than competition per se that was at the root of the problems.

However, the need to concentrate experience amongst a minimum number of institutions as in Canada and France, or at least collecting and disseminating it, as more recently in the USA, is much stronger. Also the experience of Canada and France suggests that the participation of the user in the design and construction process can be a positive factor.

Ultimately, no route to commercialisation and product development emerges as clearly optimal and in many cases the route is dictated by externalities such as a need to substitute rapidly for another fuel. Factors such as the basic quality of the designs and the personnel will inevitably be at least as significant as whether some idealised scheme of technology development is followed.

The incorporation of learning. As was argued in Chapter 2, 'learning', defined in its broadest sense of profiting from experience, has the potential to be one of the most powerful forces for technology improvement. This process is shown diagrammatically in Figure 9.1. To obtain the fullest benefits of learning, it was argued that two conditions were necessary:
– efficient communication of experience amongst interested parties
– the acquisition of operating experience with each generation of design before subsequent generation designs are ordered

If these conditions are met then the experience of operating one genera-tion of plant can be fully reflected in the design of the next generation. Nevertheless if the pace of design change is such that this cannot happen, there is still, as is shown in Figure 9.1, opportunity to profit from experience.

It may be possible to make design modifications after construction has started, and improved practice in construction and operation can certainly be incorporated in existing and later designs.

If we examine the commercial experience of nuclear power in the USA, there would appear to have been a number of barriers to efficient learning. The effect of these barriers to learning is illustrated by the operating record of reactors. There is no discernible consistent trend for existing reactors to improve through time nor is there a trend for later reactors to out-perform their predecessors.

This can partly be explained by a number of factors. First, the orders for nuclear plant that have been placed, and have not been subsequently cancelled, were placed before there was any operating experience of large (more than 1000 MW) reactors. Although some modifications could be made to the later orders in the light of the operating experience of early units, this was inevitably less satisfactory than having this experience at the start of the design process. A good illustration of this was the reduction in the volume of the reactor containment which was a feature of later orders. Operating experience showed this move to be a false economy making maintenance and repair considerably more difficult.

A second factor mitigating against efficient learning was the large number of companies and institutions involved. In the USA, most utilities use specialist architect-engineers and constructors. This is in contrast to the other three countries where these functions are often carried out by the utility or vendor. Thus, in the USA, there is an additional layer where valuable information can 'leak' out of the system.

In addition to this extra layer, in the USA there was a large number of companies that were drawn into nuclear plant supply including four vendors and 12 architect-engineers. Had the nuclear ordering rate continued at about the 20 per year that applied from 1966 to 1968, or more than the 30 per year that applied from 1971 to 1974, such a supply network might have been appropriate, but the subsequent cancellations and slump in ordering has left the experience gained very thinly spread.

At the utility level, the fragmentation of experience was even more marked. More than 50 utilities ordered plant, a significant proportion of which were not prepared for the task ahead. The utilities belatedly realised the importance of pooling their resources to maximise their knowledge and market-power through utility bodies such as EPRI set up in 1972 to research all aspects of electric utility operations and INPO set up in 1979 to promote good practice in all aspects of nuclear power. Such organisations can perform a number of functions including:
– carrying out research on common equipment problems

- establishing and promoting good practice in construction, training and operating plant
- specifying user requirements to the suppliers

The late establishment of these two organisations meant that they have been pre-occupied with the 'fire-fighting' aspects of the former two functions and are only now beginning to move towards the third function. The American ethic (often backed up by legislation) of maintaining competitive suppliers meant that none of the large utilities set up close relationships with a single supplier in the way that EdF did with Framatome in France and Ontario Hydro did with AECL in Canada.

American utilities were also slow to accept that, in most cases, operating practice at their plant was inferior to that found in other countries. The utilities were too ready to ascribe poor performance to over-severe regulation and the impact of intervenors. As the number of reactors in service outside the USA has increased in recent years, the evidence of this superiority has become overwhelming. Again, INPO and EPRI are leading the way in responding to this by increasing foreign participation in research programmes, inspection and assessment. This exchange of information and pooling of resources should be of benefit not only to the USA but also to the other participating countries.

Ontario Hydro's patronage of AECL meant that there was no sharp point of commercialisation nor a bunching of orders. Experience was highly concentrated and short construction times meant that learning could be embodied relatively quickly in new plant. There was however considerable overlap between the end of construction of the first set of orders (Pickering) and the design of the second set (Bruce). A third set of reactors (Point Lepreau, Gentilly II etc.) were designed with a little operational experience of Pickering available to work from but were not ordered by Ontario Hydro. This leaves the Darlington units as perhaps the true second generation design. A particularly salutary example of the importance of operating experience came with the fuel tube rupture at Pickering. It had required 12 years of operation to find that an equipment failure mode that had previously been thought incredible was in fact possible. Nevertheless each group of reactors entering service has tended to out-perform its predecessors suggesting that learning has been an effective process.

In France, EdF's patronage of Framatome was similar to Ontario Hydro's relationship with AECL. However, the political objectives of the programme dictated very rapid ordering and, despite very short lead-times, 27 units had been ordered before the first of them had entered service. This pace of ordering meant that few risks could be taken by making changes to the design, although Framatome was able to draw, to some extent, on

American operating experience via its licensor, Westinghouse. The first design which could draw fully on French operating experience was the N4 design first ordered in 1984. As with Canada, the limited number of companies and organisations involved has meant the experience that has been acquired should be retained and used efficiently. Evidence for this is provided by the fact that each 'tranche' of plants entering service has tended to outperform its predecessor.

FR Germany's institutional framework bears some resemblance to that of the USA with multiple utilities and a competitive reactor supply system. However, there are no specialist architect-engineers and AEG and Siemens combined their reactor interests into KWU, to give a much tighter supply side. On the utility side, there is little of the variability between utilities that is such a pronounced feature of American experience. Much of the plant is owned by consortia of utilities. This factor, combined with the existence of strong utility associations such as VGB and VDEW, may have led to a much more efficient exchange of information than occurred in the USA. In addition, many of the utilities are diversified companies such as VEBA and RWE with substantial interests outside electricity generation and perhaps greater technological strength than would be expected simply on the basis of electricity generating plant owned.

Legal challenges in FR Germany have severely stretched lead times and the change in criteria for pipe-work specification has meant that the BWR could not be commercially promoted until these problems had been demonstrably resolved. Nevertheless, the BWRs entering service now have profited from considerable operating experience and this is reflected in the excellent operating performance of these new units and the very much improved performance of the older units following the major repairs.

The technological progress of the PWR in Germany has been much smoother and the 'convoy' of plant ordered in 1982 was based on considerable operating experience of similar-sized reactors. Whilst the performance of the first large reactors (over 1000 MW) was disappointing compared to that of their smaller predecessors, it is now improving substantially and subsequent large units have shown a major improvement over the first units of this size.

Overall, the efficiency of learning appears to have been a major determinant of the performance of plant. However, whereas learning is often portrayed as a process that will naturally occur with the accumulation of experience, the evidence of the four countries studied is that successful learning is the result of purposeful actions. The obverse of this is that if learning acquired is not carefully nurtured it can be lost again. The slump in nuclear ordering means that all four countries risk this loss, particularly the USA. Skill and knowledge will be lost, first in design and manufacture, with operating skills least at risk because utilities will continue to operate existing units.

Transfer of technology. The two countries with experience of inwards transfer of technology are France and FR Germany which have both licensed American light water reactor technology, but using very different approaches. In FR Germany the two long-standing heavy electrical companies, Siemens and AEG, took out licences at a very early stage in the 1950s. Their chosen partners were companies with which they had well-established links, Siemens with Westinghouse and AEG with General Electric. Both German companies chose to make important changes to the US designs, with AEG's changes being the more radical. As was noted in the chapter on FR Germany, the changes to the PWR were more successful and it is now arguable that, on the basis of operating performance, the KWU PWR is technically superior to the Westinghouse parent design. The errors made by AEG were very costly, although recent evidence of the new reactors entering service is that the KWU BWR may now also be technically mature.

Both AEG and Siemens claimed that, by the time of the amalgamation of their reactor interests in 1969, they were independent of their licensors (in fact the Siemens licence ended formally in 1969 and the AEG licence ended somewhat later).

Although Framatome had held a licence from Westinghouse since 1958 and had participated in the Chooz reactor ordered in 1960, development work was inhibited by the French Government's support for gas-cooled technology. The standardised programme of PWRs departed little from the Westinghouse design although the new N4 design is much more distinctive. Even after the ending of the licence with Westinghouse in 1982, Framatome has chosen to maintain close co-operative links with this company.

There were a number of important reasons for the contrasting approaches adopted in France and FR Germany. First, the developments in FR Germany took place during the 1960s before the climate towards nuclear power had changed. There was still confidence in the ability of designers to find solutions to any problems they might face. By the 1970s, the climate had become much more cautious.

Second, the two German companies had always been near or at the technological forefront of heavy electrical engineering and were confident of their innovative ability. French heavy electrical engineering had been fragmented and weak up to the time of the major rationalisations.

Third, the French programme was of such scale that it demanded that no further unnecessary risks be incurred. Ordering a very large number of unproven reactors would not have been an acceptable risk. By contrast,

248 The realities of nuclear power

orders for nuclear plant in FR Germany were never a major element of government policy and were left largely in the hands of the utilities who did not have the ability to finance a similar rate of ordering to the French, even if they had wanted to.

Overall it is not sensible to try to evaluate which policy was the more successful as the situation of the two countries was so different. Nevertheless, their experience does show that radical change to a reasonably proven design represents a substantial risk. Whether this risk is worth taking will depend on the capability of the company making the changes and the scale of the benefits available. These benefits are not only in improved performance but also in enhanced competitiveness on the world market. To set against these benefits, if the licensee does not develop a distinctive product, it may prove difficult, as Framatome is now finding, to win a large share of the very limited reactor export market.

Standardisation. Standardisation has been widely seen as a solution to some of the problems of nuclear power. In particular it is expected that standardisation would:
– yield economies in manufacture
– increase the quality of the balance of plant
– reduce regulatory procedures

However, as was argued in Chapter 2, standardisation can be used to mean a variety of concepts and rather tighter definition is required for particular proposals to be evaluated. In Table 9.1, a number of standardisation schemes are listed (all of which were discussed in detail in the case study chapters), in ascending order of rigour of standardisation. Alongside each, the particular pros and cons of each scheme are summarised.

In general greater standardisation should increase economies of scale, reduce regulatory and design work and increase learning whilst the disadvantages lie in the risk of standardised error and technological stagnation. How utilities balance these points depends on their own perceptions and priorities – EdF clearly believed that the technology was mature, and their priority was rapid implementation with short construction times and low costs. Given these perceptions, the highest degree of standardisation was necessary, although it will be some time before it can be judged whether the risks were justified.

One positive advantage of standardisation, to whatever degree, is that it should ensure that the design is fully specified and agreed with the safety regulators prior to start of construction. There is increasing evidence in recent years that this is an important factor in keeping construction times and costs under control. Standardisation, almost certainly not as rigorous as that adopted by the French, may well prove useful in containing costs

Table 9.1. *Advantages and disadvantages of differing degrees of standardisation*

Example	Basic feature	Advantages	Disadvantages
1. EPRI research programme	Defined performance of each subsystem	Helps ensure good design standards are always achieved	May encourage meeting the standard rather than striving for the best achievable
2. Palo Verde project	Uses the standardised balance of plant layout employed with SNUPPS but a different NSSS	Use of proven layout reduces unforeseen logistical problems during construction and reduces resource requirements for design and licensing	
3. Germany 'convoy'	Standardised plant layout and components	If a technical problem arises, facilitates investigation of the number of reactors affected and minimises the number of repair procedures needed	
4. Overall French	Single sourcing of standardised components	Funds and expertise can be concentrated on single centres of manufacturing	Obligation to use the supplier and maintain a flow of orders
5. SNUPPS, 'tranches' within the French programme	Bulk purchase of standardised components and standardised design and layout		Risk of a manufacturing error affecting many units

and construction times and simplifying licensing procedures, but it is far from being a cure-all for the problems that nuclear power is suffering. The solutions to these lie more in the management and organisation of the technology.

Economies of scale. One of the most powerful forces shaping nuclear power development has been the variety of measures, including increased unit size and adoption of mass-production techniques, that can be termed economies of scale. A basic problem in discussing economies of scale is the difficulty of demonstrating unequivocally how large the effects are and, even, whether they exist at all. The main reason for this is that there are many important influences on economic performance of which scale is only one. Others include the quality of the utility and architect-engineering and the date of commissioning.* Unless these other factors are constant between the two samples compared, it will be difficult to establish the extent to which any differences found are attributable to scale effects or to other variables.

Nevertheless two particular examples, the rapid scaling up of American unit sizes in the mid-1960s and the large standardised French programme of PWRs, are worth examining. The rapid scaling-up of unit sizes in the USA was an attempt to gain scale economies to improve nuclear power's competitiveness against fossil-fired plant.

In the event, it has proved impossible to demonstrate whether any economies (or diseconomies) of scale existed. The reasons for this are discussed in detail in the Chapters 4 and 5. The main points are that specific capital costs (i.e. per kilowatt installed) have not been demonstrably lower and the operating performance may have been poorer for large units. As was stated above, neither effect can be demonstrated unequivocally, but what can be concluded is that scale economies were not substantial and were not automatically won. There is evidence from FR Germany in the disappointing performance of the first large units that large units are intrinsically more difficult to build and operate, but the outstanding performance of the latest large units suggests that these difficulties can be overcome.

The other major point in favour of large units is that they reduce the number of sites and planning procedures required to attain a given capacity. Against this, smaller units are likely to be easier to finance and may match better the relatively slow growth in electricity demand that is now widely anticipated. For the future, although there remains a strong belief in industry that scale economies for building larger units do exist,

---

* For example, this will determine the extent of safety systems required which will in turn be reflected in the capital costs.

such economies will form a much lesser part of the decision on what size of unit should be ordered.

The French standardised programme of PWRs illustrates a different type of scale economy – economy of number rather than size. These economies were expected to arise from a number of sources including use of efficient production-line techniques and the spreading of a wide range of overhead costs more thinly. Overall it is clear that this strategy has resulted in impressively low costs although there are factors which may reduce this advantage. First, the supply infrastructure set up is very rigid requiring a large and steady flow of orders which it is now clear is not forthcoming. The costs of either maintaining the capability in some way or scaling it down to one nearer current needs will be very high, involving severe regional unemployment and disruption. Second, the risks of standardisation (discussed elsewhere) are real and will continue to be important until the design is fully proved (if that is indeed possible) or the proportion of French electrical capacity represented by these units falls significantly.

Economies of size were also sought in the French programme by increasing unit size from about 900 MW to 1300 MW. As with the USA, economies of size have not been detectable although changes in the design requirements such as changing to single autonomous reactors rather than pairs of partially interdependent reactors may have led to any economies being obscured.

Investment appraisal techniques

The record of investment appraisals in nuclear power is a poor one. The estimates of construction costs, construction time, operating performance and electricity demand have almost invariably been very inaccurate. With a new technology that is not firmly established, it is perhaps understandable that estimates of its performance tend to err on the optimistic side. However, the degree of over-optimism that has occurred suggests that the normal enthusiasm of the promoters of the technology was not being balanced by what should have been the dispassionate view of the utilities.

There seems to have been a number of factors underlying this failure amongst utilities. These include:
- government backing and endorsement of the technology
- a relish for prestigious high-technology projects
- effective salesmanship by the reactor vendors

Whatever the explanation, the overselling of nuclear power has not served it well, creating expectations which could not be met and contributing to the back-lash in public opinion against nuclear power.

The factors underlying the overestimation of electricity demand are somewhat different. It was probably due in part to the forces inherent in almost any organisation which tend to make it try to increase its size. More concretely, most utilities, particularly since the Second World War, had been struggling to keep up with demand growth and to modernise their electricity supply systems. In this situation there appeared to be little reason to examine determinants of demand too closely, and it was assumed that historic rates of demand growth would apply for some time to come.

For the future, there will be little excuse for such over-optimism given the substantial quantity of experience which can be utilised. Of course, as with all large projects, there will be a political element in any decisions on nuclear plant and if the political will is strongly in favour, the results of investment appraisals will not be central to the decision-making process.

Turning to the investment appraisal techniques themselves, the main problem that has emerged is the use of discount rates, particularly over the very long term. The issues concerned are argued in some detail in Chapter 3. The fundamental problem is that discounting was never conceived as a technique that would be appropriate over several decades or even centuries and many of the assumptions on which discounting is based appear questionable over a long period. In particular there are fears that long-term discounting may store up debts for future generations, which will be difficult to meet because economic growth falls short of the levels implicitly assumed in the discount rate chosen. There is also dispute about the level at which the discount rate should be set, whether it should act as a capital rationing device and how provision for future debts should be made. If future nuclear investment decisions are to command confidence that adequate consideration has been given to future problems and costs, a necessary, but not sufficient condition, may be that these issues are subject to a more explicit and open debate than has so far been the case. One possible outcome of such a debate might be that costs incurred beyond the productive life of the reactor should be discounted at a low or even zero rate in the period after the shutdown of the reactor.

Institutional structure

Emerging from this study, it seems that the key institutions are the utility, the vendor and the safety regulator, with the influence of government also significant. As was argued, the way in which these institutions are constituted and have evolved is heavily dependent on the economic history and culture of the country and there may be comparatively little freedom to change them. The object of this section is therefore not to prescribe a particular structure but to highlight the key responsibilities and functions.

The utility. All the evidence in this book points to the centrality of the utility in determining the success of nuclear power plant. This is partly because, ultimately, the utility must bear the responsibility for choosing the technology and building and operating the plant. It is also because, whereas with many technologies, operating or 'using' them is a relatively simple process, 'using' nuclear power plant is clearly a very demanding task with ample scope for failure if insufficient care is taken.

If utilities are crudely categorised according to their success, the most unequivocally successful are probably Ontario Hydro and some American utilities. EdF and the German utilities can also be seen as relatively successful whilst other American utilities have the poorest record. Using the simple characteristics of the utilities, it is hard to identify any common features between the successful utilities. However, it is worth noting that some of the factors which might have been expected to be important appear not to have been. These include:

- size of utility. It might be expected that large utilities with strong formalised management and a powerful independent engineering capability would perform best with nuclear plant. In fact although some large utilities have performed well, such as Ontario Hydro, of the largest US utilities, only Duke Power's record is much above average.
- architect-engineering and construction. It might be expected that those utilities which carry out their own architect-engineering and construction would be better motivated and would produce designs which closely matched their own requirements. Again, the records of Duke Power and Ontario Hydro support this hypothesis, but the records of the Tennessee Valley Authority and Pacific Gas & Electric contradict it
- ownership. There appears to be no systematic relationship between a utility's performance and whether it is publicly or privately owned

In fact, closer examination of the successful utilities does reveal some rather more subtle common features and these can be summarised as management strength and a critical technological capability.

Management strength is required at a number of stages. Prior to construction the utility must select which activities it should carry out itself and which it should contract out. Where the utility contracts out, it must not only select a competent company but ensure that the company chosen assigns a good team to the job. During construction, the utility must ensure high quality work and tight control of schedules and expenditure. It must also ensure close contact with the safety regulators to avoid conflicts and costly design changes. During operation, the utility must ensure the good training and motivation of the operators and maintenance staff.

Technological capability, particularly being an informed buyer, is also important at all stages in the product cycle. Prior to ordering, the utility must have sufficient technological capability to select a sound technology. Prior to construction the utility should be able to contribute to the detailed design ensuring the incorporation of features which will allow it to operate and maintain the plant efficiently. During the operation phase the utility must be able to make strategic decisions about when repairs should be carried out and whether faulty systems should be replaced or repaired.

None of these management and technological criteria require the utility to be particularly large although if the utility is to make a major contribution to the direction of technological change, as Ontario Hydro has done with the CANDU reactor, this is only likely to be possible for a large utility.

The vendor. In the framework set up in Chapter 2, a number of issues regarding the degree of integration and diversification and ownership were identified. As with utilities it is possible to make a crude categorisation according to degree of success with AECL and KWU (Siemens) at the top, Framatome and Combustion Engineering in the middle and KWU (AEG) and the other three US vendors at the bottom. Again, as with utilities there are no simple formulae for success – AECL and Kraftwerk Union have little in common with each other in terms of ownership, origins and diversification although they do both sub-contract most of the manufacturing. Thus it is neither possible nor desirable to try to lay down a blue-print but it is possible to identify risks and drawbacks given a certain approach.

The two extremes of horizontal integration and adoption of mass-production methods are represented by Framatome and AECL. The high degree of horizontal integration of Framatome appears to have been an important factor in maintaining the smooth schedule, low costs and consistent quality of the French programme. However, the current and continuing turndown in ordering rate illustrates the problems with this degree of rigidity. Even though it has been clear since the early 1980s that no more reactor orders were needed for some time, the French Government has continued to allow and even encourage the placing of orders to ensure that production lines are kept loaded, resources and skills are not lost, and that Framatome remains reasonably viable as a company.

On the other hand, AECL and KWU which have subcontracted component supply and whose ordering rate has warranted, at best, batch production methods, may have paid some financial penalty by not getting components as cheaply as Framatome. However, they have been able to select suppliers on the basis of quality and cost, and in periods of little or no ordering, the burden is shared with their supply network.

The American vendors all set up substantial component supply facilities in the early 1970s but, with the famine of orders of the last 10 years, these have now largely been shut down. However, the American vendors have all been wary of becoming involved in the overall design of the plant and, given that many of the problems nuclear power has experienced were connected with the balance of plant, this horizontal integration may not have been the best use of their resources.

Turning to the degree of diversification and ownership, as was suggested in Chapter 2, product diversification and public ownership can provide a degree of protection against adverse market conditions. This is illustrated by Kraftwerk Union in the former case and Framatome in the latter. However, this protection is only limited. General Electric, for whom nuclear power represents only a relatively small proportion of their business, are now showing little interest in competing vigorously for new orders. If a substantial volume of business became available, GE would undoubtedly step back into the market, but GE would probably not be greatly concerned if their nuclear reactor supply capability was lost.

The Canadian Government's support for AECL must also be finite and, unless new orders are won soon, it is difficult to see how a reactor supply capability can be maintained. A flow of business in reactor servicing will inevitably continue but this may be better met by absorbing parts of AECL into Ontario Hydro than maintaining AECL as a separate entity.

The safety regulator. Perhaps one of the most important requirements for a successful nuclear power programme is for a strong, demonstrably independent safety regulator. For nuclear power to be publicly acceptable, safety regulators must be seen to be applying stringent standards to the utilities and coming down hard on the utility if standards slip. However, for various reasons, none of the four countries has achieved this goal and regulators are frequently seen by the most committed supporters and opponents of nuclear power as being captive to the other side. The fact that safety regulation is the responsibility of government means that it will inevitably be difficult to convince sceptics that safety regulation is not influenced by government policy on nuclear power.

In the USA the task of the NRC has become almost impossible. It inherited a vast backlog of orders from its predecessor, the AEC, which were placed before many of the fundamental safety issues, such as protection against loss of coolant accidents, had been thought through. In addition nearly all of these orders were different and many of the detailed designs of the balance of plant were ill-conceived. Their major problem however was the poor standards of a few of the worst-managed utilities. Inevitably rules had to be framed which were aimed at these poorest utilities which led to a proliferation of new regulations disrupting the operation of even the best-run utilities. The NRC can perhaps be blamed for not tackling the root cause of the problem, the inadequate standards in some utilities, but US utilities seem rather resistant to external advice. Perhaps the best hope is the utility association, INPO, which has made impressive progress in raising general standards. If INPO can reach some form of accommodation with the NRC to avoid duplication of efforts and for NRC to recognise more formally INPO programmes, and if INPO's efforts are effective for *all* utilities there is some prospect of an improvement in the regulatory system.

In FR Germany, regulators faced pressure from opposition no less determined than in the USA. However, there is much less evidence of poor utility standards than in the USA. Two factors have complicated the German regulatory system, the devolution of safety regulation to the *Länder* and the use of the courts to resolve disputes. Logically there can be little justification for requirements to vary from Land to Land, but the autonomy of the *Länder* is jealously guarded and the federal government is very wary of overriding local government decisions. Although the 'convoy' scheme has had some success in establishing national standards, it remains to be seen whether these standards will be accepted in *Länder* less well disposed towards nuclear power. Whether the courts are the appropriate place to resolve technical disputes is arguable, but the process may have served the purpose of defusing, at least temporarily, the violent protests that had become common place in the mid-1970s.

In both FR Germany and the USA, utilities have claimed that regulation has been too stringent and that safety requirements are too strict. Whether the additional safety requirements have made a worthwhile contribution to safety is very difficult to answer. However, a number of points can be made. Firstly, changes in safety requirements can be extremely disruptive to costs and schedules but where designs have been fully worked out and discussed with the regulators prior to the start of construction these disruptions can be minimised. Secondly, in both the USA and FR Germany the tightening of regulations has followed incidents which have undermined confidence in the abilities of the utilities. In the USA these were the poor operational practices typified by the Browns Ferry and Three Mile Island accidents and in Germany, these were the quality and design problems associated with the BWR–69s.

France and Canada both have the feature of very close relationships between utility and regulator. This undoubtedly produces efficient communications and a high level of trust between utility and regulator. However, this closed, relatively informal style of regulation may not command public confidence that it is independent and rigorous. In Canada, this problem is recognised and the AECB is taking steps to open its proceedings to public scrutiny. Whether this can be accomplished without causing too severe disruption to the process remains to be seen.

In France, there seems to be little recognition of this problem and the belated disclosure of the details of incidents with significant safety implications is damaging to the credibility of regulators.

Government. The influence of government is partly felt through the policy and style of the safety regulator (discussed above). The government is responsible for determining the constitution of the regulatory authorities, the resources allocated and reporting procedures for the safety of reactors. Once these are established it may be quite difficult to change them. This is demonstrated by the US NRC, which has maintained its basic structure despite widespread dissatisfaction with aspects of its operation.

Equally important is the government's policy on finance and siting procedures. This is well illustrated by the contrast between France and the USA. In France, an expansion in nuclear power was a major plank in government policy. To ensure the success of this policy the French Government set down a framework encompassing the following features:
- the provision of government backed loans at preferential interest rates to EdF
- the use of planning procedures which allowed no effective scope for public opposition, backed up where necessary with strong policing
- the political will to impose drastic rationalisations on the previously fragmented heavy electrical supply sector

Without this substantial government backing, EdF would not have been able to carry through a programme anywhere near the scale that has been accomplished. Whether this policy of rapid expansion was appropriate is discussed in the chapter on France but there are additional pros and cons to the dirigiste way in which the policy has been carried out. Whilst this dirigism certainly contributed to economic and managerial efficiency, there must be concern about the degree to which the population feels it has been party to the decision-making. While things are progressing fairly smoothly as they are now, this presents no problems but if a significant accident were to occur undermining confidence in the safety of the programme, public condemnation could be much stronger than if the programme had been carried through by more open, democratic procedures.

At the other end of the scale, the American Government has adopted a very much more laissez-faire attitude, at least in financial terms. Government support through provision of fuel cycle facilities was important in the early stages but as the problems have increasingly centred on financing the reactors themselves – an area in which the government could not intervene – the government has had less and less impact on the fortunes of nuclear power. President Carter's reservations about nuclear power and President

Reagan's strong support have had little effect on the US industry because they could not be backed up financially or through influence over the utilities.

The cases of Canada and FR Germany are made more complicated by the federal structure of these countries. The CANDU is very much the product of Ontario province. The government of Ontario, to which Ontario Hydro is responsible, has shown strong support for CANDU. At the federal level, to which AECL reports, support has been less committed, although CANDU has always been treated sympathetically. For the future, with the fate of CANDU increasingly dependent on export sales of reactors, the stance adopted by the federal government in facilitating finance will be increasingly crucial.

In FR Germany the federal/regional tensions are rather different. Some of the *Länder* are sceptical about nuclear power whilst some are hostile to it. The federal government, which has always supported nuclear power strongly, for example through research centres, has been unwilling and sometimes unable to override local decisions leading to a very uneven geographical distribution of plant.

Overall, it must be concluded that, in terms of short-term economic efficiency, a centralised, authoritarian approach is most effective. However, such an approach is only likely to be possible in countries which already have a politically centralised system. In addition, although centralisation does generally lead to managerial efficiency, there is no evidence that technical problems are more effectively dealt with by such a system, and for the longer term, a lack of democratic accountability in decision making may rebound on the policy.

CHAPTER 10

The future of nuclear power

Introduction

Over the past ten years, many of the features that characterised and underpinned the previous development of nuclear power have changed and the first era has clearly ended. Some of the factors behind this, such as the Three Mile Island and Chernobyl accidents, were highly visible discrete events whilst others, perhaps of equal significance, such as the failure to control capital costs and the long-term reduction in electricity demand growth, have been much more gradual and pervasive in their effect.

In this chapter we look beyond the four case study countries at the current status of nuclear power worldwide. In particular, the various perceptions of nuclear power are examined, including those of governments, utilities, vendors, the public and, of increasing importance, the financial community. The new era for nuclear power is then considered, particularly the factors that will shape it, including the demand for plant, the choice of technology, the structure of industry and the various challenges, new and existing, that will face the nuclear power industry.

The current status of nuclear power

Nuclear power has never been, and never will be, just another energy technology. Its military connotations, its potential risk to public health and to the environment, and its extraordinary requirements in terms of financial, technical and human resources have always meant that decisions on nuclear power are taken at a high level and often hinge on factors other than simple cost minimisation criteria. Many decision-makers have ascribed to nuclear power a non-cost advantage over competing technologies on the basis of factors such as:
– its military connections
– scientific prestige and its high technology content

– the security it appeared to offer against price increases and interruption
to fossil-fuel supplies
– the development of skills that would be increasingly in demand in the
  future and which would lead to increased national success in world trade

These factors led governments to sponsor research and development and
even subsidise apparently commercial decisions on the purchase of generating plant. However, many of these non-cost factors are now in the process
of being re-assessed.

The promotion of civil nuclear energy as an alternative to nuclear
technology's military applications meant that the dual-function nature of
nuclear installations such as enrichment plant, reprocessing plant and even
reactors could enjoy considerable cross-subsidisation of the civil aspects by
the less cost-conscious military aspects. Increasingly this degree of
ambiguity about the purpose of nuclear installations has become unacceptable, and the military and civil sides are being more strictly separated.

Nuclear power appeared to be the next in a line of technological
innovations making electricity generation more sophisticated and 'high-
technology'. There was a common belief that nuclear power would inevitably become the major method of electricity generation in all countries and,
further, that electricity would increasingly replace fossil-fuels for final
energy use. However, worldwide energy demand growth has fallen drastically over the past ten years,* and this factor along with substantial new
discoveries of coal and gas has meant that fossil-fuel reserves are now
expected to last very much longer than they were at the time of the first oil
crisis. (For an account of the prospects for gas and coal, see International
Energy Agency (1986a and b).) The further scope for cost-effective energy
conservation is considerable, especially if fossil-fuel prices increase from
their current low level. In addition, the world market for nuclear equipment
and know-how has proved small and the few countries that have built up a
capability have found the world trade that has accrued to be very lucrative.

Nuclear power also appeared to be a less economically risky investment
than, for example, coal-fired plant. Completing plant to time and cost was
not perceived to be a major problem in the early part of nuclear power's
commercial life. Nuclear power appeared to be a largely risk-free investment once it was completed, with little vulnerability to rising fuel costs or
interruptions to fuel supplies. It is now clear that in most countries,
completing plant to time and cost is not a certainty, and operating
performance is not guaranteed to be good. Thus, the capital charges, which

---

* In the period from 1974 to 1983, world primary energy production grew at an average of less
than 1% per annum. This compares with an average growth rate of nearly 5% in the period
from 1950 to 1974.

are usually expected to represent about 70% of the unit cost of nuclear generated electricity, and the number of output units over which these charges can be spread are difficult to predict. The risk of serious disabling accidents which have repercussions for other reactors in terms of plant modifications has been illustrated dramatically by the accidents at TMI and Chernobyl.

Despite all these changes, utilities are still, on the whole, enthusiastic supporters of nuclear power. For many utilities, particularly those which completed plant before costs began to escalate rapidly, nuclear power has proved a reliable, cheap source of power because of its relatively low running costs. Other utilities with a less happy experience of nuclear power have tended to blame external factors such as public opposition and unsympathetic, over-stringent regulation rather than failings of their own or of the technology. The belief still seems strong that once this 'interference' is removed, nuclear power will become a cheap source of power entirely under the control of the utilities, unlike fossil-fired plant where the utility is inevitably at the mercy of the fuel supplier.

The vendors have also maintained their overt support for nuclear power and, perhaps not surprisingly, seldom acknowledge past failings. However, closer scrutiny of the vendors' strategies suggests that their perceptions are changing. In the past, particularly for those companies for which nuclear power is only a small part of their business, nuclear power was regarded as a strategic capability which would be of growing importance and which should be maintained in spite of short-term losses. However, increasingly vendors are now orienting their efforts towards areas such as reactor servicing which offer a more secure business opportunity and nuclear power is a declining proportion of their business.

Public responses to nuclear power are complex and difficult to define. Nevertheless a number of factors have emerged or been confirmed in the last few years:
- public distrust of statements from all parts of the nuclear business including governments, utilities, vendors, international agencies and atomic energy research establishments has become increasingly strong
- the hope that public fears of nuclear power would recede with better education and greater familiarity with the risks has proved unfounded. The strength of opposition may ebb and flow but it is unlikely to go away

For an increasing number of countries, the future of nuclear power is in doubt. Previously many people had assumed that nuclear power was inevitable and that it was only a question of timing as to when the option would be taken up. However, a number of developed countries including Sweden and Austria have taken decisions not to use or to phase out nuclear power whilst other countries such as Switzerland, Spain, Holland and perhaps even FR Germany, the USA and the UK may decide overtly or by default not to expand capacity. Whereas in the past, in most countries there was broad political support for nuclear power, in countries such as the UK, FR Germany and Spain, nuclear power has become a party political issue and the perception is growing that there is little to be gained politically by supporting nuclear power and much to be lost. In developing countries, problems of finance have halted the expansion of nuclear power and, particularly for the poorer developing countries, nuclear power may now be seen as an unjustified diversion of human and financial resources rather than as a positive contribution towards development.

The financial community's influence over nuclear power has only recently been recognised but is of considerable importance. As unit costs rose, it began to be recognised that the utility's ability to obtain finance could place an upper limit on the size of programme that could be undertaken. However, since the TMI accident and the increasing recognition of the risk of long lead-times and cost over-runs, financing even one reactor has become a major problem for almost all utilities except those with the firm backing of the government of a large strong country. This factor would in itself probably be sufficient to prevent new reactor orders in the USA for some time. It has also led to a number of developing countries such as Egypt, Turkey and Mexico going through an elaborate evaluation of bids for reactor sales, only to find that finance was not available.

The new era for nuclear power

The future demand for plant. Much of the commercial history of nuclear power has taken place against a background of steadily decreasing demand for plant as expectations of future electricity demand growth have been revised down. Even the market for power stations to replace old fossil-fired plant, which was expected to become increasingly important, is under threat from the increasing attraction of refurbishing existing stations.

There is still considerable disagreement about the future levels of electricity demand growth. Those that anticipate high demand growth (3–4% per annum) suggest that electricity is a uniquely high grade form of energy that will continue to find new uses. They point to the relatively high rates of electricity demand growth that have occurred since the world recession of the early 1980s and the increasing use of micro-electronics. They also point to the close correlation between economic and electricity demand growth and they suggest that the economy cannot expand at a healthy rate without rapid electricity demand growth. Those that expect low demand growth (less than 2%) would suggest that recent increases in demand are a short-term reflection of the recovery from recession, and whilst new applications of electrical technology will be numerous, they will use comparatively little power. They would claim that many of the factors that brought about the reduction in demand growth will continue to dominate. In developed countries the main factors have been:

- saturation in the ownership levels of many of the most important electricity-using appliances in households
- a relative (and in some cases absolute) decline in the most electric-intensive sectors of industry
- rapid and continuing improvements in the efficiency with which electricity is used in all sectors

In developing countries, the situation is rather different. Whilst some countries such as South Korea and Taiwan have experienced high economic growth leading to comparably high rates of electricity demand growth, other developing countries, particularly those in Africa, have not even kept pace with the developed countries and their electricity demands remain a very small proportion of world demand. Undesirable as this situation is, there is little optimism that things will change rapidly – the current low commodity prices are likely to make things worse. Even if this situation were to change and developing countries did begin to make rapid progress in catching up with the developed countries, nuclear power might be slow to profit. The main reasons are:

- few developing countries' electricity grids are large enough to efficiently accommodate a nuclear reactor
- other projects with an apparently stronger contribution to make towards economic development will have a prior call on the limited financial and human resources

This situation might change if cheap small reactors (about 250 MW) became commercially available but, as is argued in the next section, this must be regarded as unlikely.

For most developed countries, it seems likely that electricity demand will not grow at a long-term rate of much more than 2% per annum and short-term surges in demand growth will be balanced by periods when demand growth is low or even negative. Given that, particularly in the USA, the risk of over-ordering plant, both in economic terms and from public condemnation, is now much greater than it has been, few utilities will have the confidence to embark on a major plant ordering programme. For many utilities, a 1000 MW plant will represent several years of demand growth. This plant demand scenario has a number of consequences.

Firstly, very few countries will experience sufficient electricity demand growth to sustain an independent indigenous nuclear power plant supply industry. Companies in smaller countries will tend increasingly to seek international alliances and mergers. Such alliances will largely overcome the apparent stigma of having to rely on imported technology but will effectively further narrow the range of technological diversity.

Secondly, with few orders being placed, the degree of success of each plant will be conspicuous and few utilities will wish to take the risk of ordering unproven plant, particularly new reactor systems but also new designs of existing types.

Thirdly, for some utilities, measures such as electricity conservation, load management and encouraging industrial co-generation capacity, which are less economically risky to the utility than ordering new plant, will be an increasingly attractive option.

Unless there is a drastic improvement in one or more of the aspects of this scenario, the only non-communist countries where further orders for plant can be confidently expected are France and Japan. In these countries centralised planning, the absence of substantial indigenous fuel resources and strong government support can be expected to continue to overcome difficulties. These two markets might generate a total of one order a year over the next 10 years with most of these orders being placed in Japan. Of the other potential markets, perhaps the best prospects for nuclear power are the UK, South Korea and Taiwan. Orders for these three markets, which will probably have a substantial import content, are unlikely to exceed a total of five units over the next 10 years. Allowing for perhaps one or two orders from an unanticipated source, this gives a total of about 15 units over the next decade with no vendor winning more than about three orders.

Technology choice

There have been, and will continue to be, suggestions that a partial solution to nuclear power's problems is to be found in new, safer designs of reactor. Following the accidents at TMI and at Chernobyl, there is a perception that new designs of reactor are required which respond slowly to fault conditions and which are more forgiving of operator errors. In addition, there is now a strong feeling that the LWR designs, particularly as they have evolved in the USA, need to be fundamentally reviewed. It is felt that many of the components are redundant and that the safety systems installed do not represent the most cost-effective method of maximising safety.

However, the demand scenario outlined above, the loss of nuclear power's special status and the direction of public-funding for new reactors, particularly its bias towards fast reactors, make it unlikely that a substantially new design will be brought to a commercial stage in the next decade or two.

It is ironical that much of the present situation with respect to reactor choice and reactor development has been shaped by a reactor system which has never been ordered as a commercial choice. From the outset, the fast reactor has been seen, almost universally within the nuclear industry, as the culmination of fission reactor development with, perhaps, fusion power as the ultimate solution to energy supply. This was based on the premise that supplies of uranium would, in a matter of one or two decades, become severely depleted and a more efficient method of utilisation of uranium would be required. This dictated that reprocessing of spent fuel (a technology already developed for military purposes) to recover the unused fissile material was required and that breeder reactors which could use nearly all the uranium should be developed. Plutonium arising from the reprocessing of fuel from thermal reactors was assigned a significant positive value, much improving the economics of the fuel cycle for thermal reactors. Government spending on nuclear research reflected these perceptions and countries such as the USA, the USSR, France, FR Germany, the UK, Japan, Italy and India all attempted to develop an independent capability in fast reactors and reprocessing plant.

A combination of specific circumstances led to a sharp rise in the price of uranium in the early 1970s and, in many government quarters, this reinforced belief in the appropriateness of this line of technology development. However, by then, nuclear technology in general was not developing as smoothly as expected with costs much higher than forecast and considerably greater technical difficulties than anticipated. Fast reactor development suffered from similar problems considerably slowing development. The peak in uranium prices was quickly passed and it became increasingly clear that uranium resources were vastly more abundant than previously estimated and that nuclear expansion would be much slower than forecast placing less demands on these resources. At the same time government funding of nuclear research has generally been falling in real terms and breeder research has continued to take a large slice of the available funds despite the fact that even France, the leader in breeder technology development, does not expect to build a breeder reactor on a commercially viable basis for the foreseeable future.

In addition, there must be strong doubts as to whether the breeder would represent an acceptable and competitive technology in many countries even if the technology can be proven and demonstrated to be acceptably safe. The main reasons are:
- a technology which required the production, separation, transport and use of plutonium on a huge scale would provoke more united and determined opposition to its deployment than any thermal reactor
- the cost of uranium is not likely to rise sufficiently in the foreseeable future

to counterbalance the effect of the capital costs of breeder which are acknowledged to be higher than for thermal reactors

The pressures to reduce government spending on nuclear power are not likely to diminish and this means that new reactor development will receive little help from government sources and that, whilst the breeder will continue to be a political issue, it will not be a serious option for commercial reactor orders. The LWR is almost certain to continue to have its supporters and any new design could only hope to slowly increase its share of new orders. The main candidates for this niche include the CANDU which is already commercially available, the high-temperature reactor which has been built at demonstration plants, the Swedish PIUS system which is still at the design stage, or a totally new design. Of these, the CANDU, because it is commercially proven deserves careful attention. However, of the other options, whatever their technical merits, there are strong political and institutional reasons why they are unlikely to find acceptance. Whilst ordering of nuclear plant in the USA effectively ended more than 10 years ago and it is now arguable that the American reactor supply capability has substantially diminished, it remains true that most of the world still looks to the USA for technological leadership. Thus a new reactor system might well need the American seal of approval if it is to have credibility and this effectively requires reactors to be ordered for installation in the USA. It is therefore worth reiterating the arguments about the likelihood of further reactor orders in the USA.

At present, it is difficult to see how further orders for nuclear plant could be possible in the US. However, in perhaps 10 years' time, given continued electricity demand growth and a consequent shortfall of plant, a substantial up-turn in fossil-fuel prices, and no further accidents at nuclear power plants, nuclear power may begin to look attractive again. Even in these circumstances there are reasons why a new reactor system is likely to be chosen. First, building an LWR already represents a substantial financial risk. This arises from uncertainty about (i) safety requirements; (ii) the difficulty of controlling costs and construction times; (iii) whether the financial regulators will allow expenses to be fully recovered and (iv) the operating performance of the plant. Ordering a new unproven reactor system would add to the uncertainty on safety regulations since no precedent would exist for licensing such a plant. The operating performance of the plant would be no more predictable. Taking such an additional risk would probably not be feasible for one utility alone given its dependence on external financing, although a project with utilities sharing the risk under the auspices of EPRI may be possible. Given the poor return on the money invested in the Clinch River Breeder project and scepticism, even hostility, to the use of further federal funds in reactor development, it seems likely that the most a government favourably disposed towards nuclear power could promise would be a streamlining of safety regulation.

Second, it is difficult to see how the reactor supply industry for a new system could be built up especially given limited federal backing. The supplier would probably need to be large with sufficient resources to carry out R&D and sustain financial losses on this operation for a number of years. It would also probably need a good knowledge of, and contacts in, the utility business. Few companies which meet these criteria are not already involved in reactor supply and those already involved would be unlikely to want to repeat their painful experience with another design.

Looking to possible avenues outside the USA, two areas, Europe and Japan, would appear to have the technological strength to pioneer a new design. A European collaboration is unlikely to be possible given the prior commitment to LWRs, especially in France. Of the individual countries, FR Germany has the technological strength to carry out development independently, perhaps on the high-temperature reactor. However, unless the LWR was heavily discredited, a circumstance that would probably make any other reactor system infeasible, it is difficult to see why utilities would forego the proven merits of the LWR, which, at least in FR Germany, has been increasingly successful, for the risks of an unproven system. Similarly Japan, despite some work on indigenous designs such as the breeder and advanced thermal reactors, has shown little appetite for going it alone.

This leaves the CANDU. It has the merit of being demonstrably developed and proven. However, Canada has little history of developing and supplying high-technology products and many countries may feel wary about committing themselves to a design which has not been endorsed by a major industrialised country. Given that it has been commercially demonstrated and available for 15 years, and rejected in most countries in favour of the LWR, it is difficult to see the circumstances which would cause these countries to go back on previous decisions.

By default, this leaves the LWR and on current trends, the PWR rather than the BWR. An early priority, particularly in the USA, is a thorough review of design. The designs have been modified in order to incorporate new safety systems to such an extent that there is now thought to be a great deal of unnecessary redundancy. What is important is a swift resolution of the unresolved 'generic issues' since these may have a significant effect on the safety requirements. Once these are settled, the design can be thoroughly reviewed to ensure that the required safety standard can be met most cost-effectively.

Two routes may help this process. EPRI is currently carrying out a review of safety systems which should resolve many of these problems and both GE and Westinghouse are collaborating with their Japanese licensees in a programme to develop advanced BWRs and PWRs respectively which should also meet this concern.

Outside the USA, the need to review the design is likely to be less pressing as the current ones or those now under development in Japan (mentioned above), France and FR Germany were all conceived in the light of considerable operating experience and have not needed the amount of revision that the American designs have undergone.

Small units. In recent years there has been publicity about the possible use of small reactors (200–500 MW) in developing countries and the IAEA, in particular, has been enthusiastic about the prospects for such sales suggesting that 25–30 countries might be potential buyers before the end of this century. (For a brief summary of the argument see Schmidt, 1984, pp. 29–35.) Most vendors now claim to have designs of small reactors available, although it is not clear how fully these designs have been worked out. The main features most of the designs are said to embody are:
- a relatively 'untuned' design, such that it is fairly forgiving of operator errors
- a high level of prefabrication which does not over-burden the engineering skills of the installing country

However, despite this marketing effort, no developing country has come close to ordering such a unit and, on present trends, none is likely to over the next decade. The main reasons are:
- the designs are unproven and will appear to be economically risky unless a vendor can persuade a developed country to host a demonstration plant
- the economics of large nuclear power plant in experienced countries using established designs, are little better than those of coal-fired plant. It does not seem likely that unproven designs in inexperienced countries could be more economic than coal-fired plant
- small reactors would still be capital-intensive and would present more severe problems of finance than, for example, coal-fired plant
- the experience of many developing countries that have ordered nuclear plant, such as India, Pakistan, Mexico and Philippines has not been good, often involving long lead-times, escalating costs, disappointing operating performance, and an unsatisfactory relationship with the vendor

On balance, it is unlikely that such reactors will find a market and it is therefore unlikely that they will affect the direction or success of the world nuclear industry.

The structure of the nuclear industry. Over the past decade, the nuclear industries of the world have mostly been in disorderly retreat from their proud position of the early 1970s when technological optimism was high. Their expectations of future demand for plant have consistently been too high and their diagnosis of the roots of their difficulties incomplete. There is now recognition that no vendor, even Framatome, is going to win a steady flow of orders (more than one a year) and that major adjustments will be needed. A number of vendors, including General Electric and Babcock & Wilcox of the USA, and BBR of FR Germany, are already no longer actively seeking new orders whilst the remaining vendors are reducing their manufacturing capability and reallocating their resources particularly towards the reactor servicing market.

This market has a number of attractions to vendors:
- the market for reactor services is assured by the large number of operating reactors
- where it requires the solution to design problems, it will maintain and develop the sort of skills and knowledge of reactors which will allow the vendor to maintain a capability to design reactors
- reactor servicing will, by its nature, comprise a large number of relatively small contracts won by competitive tender. This may give vendors the opportunity to build contacts with a wide range of utilities in markets not previously exploited

This view of the way in which vendors will survive, some withdrawing and others restructuring and regrouping, is becoming increasingly widely held, and it probably represents the best outcome the vendors can hope for. However, consideration of the nature of the services market and the interests of utilities might suggest a more radical restructuring, with the leading utilities taking a much more active role in technology development and vendors employed much more on a sub-contract basis.

The services market will be extremely competitive because, unlike new reactor sales, the existence and size of the market are relatively easy to establish. It is unlikely that this potentially lucrative field will be left just to vendors. Utilities, who will in some cases have built up a very full understanding of nuclear plant, will see an opportunity to capitalise on these skills and increase their income. Specialist companies will be able to select a particular activity or repair and build up a capability which can be marketed world-wide. Even the vendors who have withdrawn from new reactor sales will have an opportunity to make some return on their investment in nuclear power.

For the future development of reactor technology, utilities are unlikely to be content with the passive role that has characterised the history of nuclear power. Indeed there is a growing number of examples of utilities taking a more positive role:
- in the USA, the programme of nuclear research carried out by EPRI is increasingly oriented towards specifying the design parameters for a

reactor. Carried to a logical conclusion this could culminate in the design being specified by the utility and its manufacture either in part, or as a whole put out to competitive tender

- in France, EdF has shown itself eager to take an increasing stake in Framatome. Also, a positive contribution from EdF is becoming an integral part of Framatome's export bids for services and reactors
- in Canada, Ontario Hydro has always taken a central role in decisions about the direction of technology change and, especially if no new orders are placed, it will increasingly be the centre of knowledge about the CANDU system
- in Japan, the utilities are taking an increasing role in stimulating reactor research. Much of the impetus and finance for the Advanced PWRs and Advanced BWRs which are now under development came from the utilities

In many respects such a shift in the location of reactor development away from the research agencies and the traditional vendors would seem a healthy change. It should lead to the development of products which match more closely the needs of utilities. However, it is not a guarantee of success. Utilities have, in the past, shown themselves capable of serious technological over-optimism, poor forecasting, national chauvinism and poor responsiveness to public concerns and needs.

The final area to be considered is the future of the national nuclear research and development agencies. Whilst their promotional activities appear increasingly anachronistic, and their efforts in reactor development have seldom produced commercially applicable designs, there are certain activities that few governments would feel content to leave to the market. These tend to be processes with close military or environmental connotations such as enrichment, reprocessing and waste disposal. However, there is no reason why these activities should not be carried out on an economic, un-subsidised basis, by commercially-run, but publicly-owned companies. Similarly there would seem to be no good reason why utilities should not be charged a full economic rate for regulatory services.

There may well be a case for state funding of nuclear research in the future but it must win that funding in fair competition with other energy technology applications. This point of view has become widespread in recent years, not just among opponents of nuclear power but also in government circles and among the competing fuel industries.

Future challenges and issues

Few of the circumstances that have made the development of nuclear power so problematic over the past two decades can be expected to ease. It is unlikely that public opposition will subside and the accident at Chernobyl is likely to increase the difficulties of siting nuclear plant. Future installations will need to be sited more remotely and more intense opposition from those living near to the plant will arise.

In addition to these existing difficulties, a number of new issues are becoming increasingly important, including the problems of cost containment, radio-active waste disposal and plant decommissioning. During much of the commercial history of nuclear power, the nuclear industry has sought to reduce or contain costs primarily by increasing the size of units and increasing the scale of production. Neither of these routes has been as effective as the industry had hoped, and they are unlikely to offer scope in the future. There seems to be little advantage in building even larger units and there is pressure, particularly in the USA, for a return to smaller units for reasons of easier finance and ease of absorption into relatively slowly growing electricity supply systems. The likely scale of future reactor orders means that few production-line facilities will be adequately loaded and equipment suppliers may need to return to 'one-off' or batch methods.

On the more positive side, one of the lessons to emerge from the case studies is that costs are most efficiently controlled by strong, effective utility management. If the low volume of orders means that utilities will be able to concentrate resources on very few units, this increased attention may counterbalance the lost economies of scale.

One set of circumstances that might favour nuclear power would arise if the environmental consequences of burning fossil-fuels, for example the greenhouse effect and acid rain, were perceived to be very much more serious than is currently the case. In this respect however it is worth remembering that the first oil crisis appeared, at the time, to give nuclear power a considerable advantage over its competitors. This advantage was very quickly lost to rising nuclear power costs. It may be that if nuclear power was presented with another radical improvement in its competitive position, that public pressures for further safety improvements, such as a requirement for underground siting, would again absorb that advantage.

It is not easy to be optimistic about the problems of waste disposal and reactor decommissioning. The nuclear industry seems confident that neither requirement will present insuperable technical difficulties. It claims that the main processes have all been developed at least to demonstration stage. Whether the step from demonstration to full commercial application can be accomplished with a minimum of technical difficulties and without major cost-escalation remains to be seen. However, the record of the nuclear industry in this respect gives little confidence that this will be the case, and the problems of public acceptability of waste disposal seem much more intractable even than reactor siting.

Nuclear power started life with an immense fund of good will from government and people alike. It is difficult to avoid the conclusion that that fund is now dangerously depleted and future failures will be punished increasingly severely.

APPENDIX 1: THE TECHNOLOGIES

Introduction

The objective of this appendix is not to provide a technical account of reactor technology but merely to briefly state the basic principles and to highlight some of the important features of, and differences between, the main technologies. The appendix also gives brief details of the commercial success and status of the main reactor types.

Reactor physics

The fundamental requirement for a nuclear reactor is a core of fissile material which can sustain a nuclear chain reaction. A fissile nucleus is one which can be induced to split (fission) by low energy or thermal neutrons and the only naturally occurring fissile material is uranium 235* (^{235}U). There are a large number of possible fission reactions and reaction sequences, each of which usually results in the production of two major fission fragments such as strontium or caesium nuclei and high energy, much smaller fragments, such as neutrons. The requirement for a chain reaction is that there be no *net* absorption of neutrons – some of the reactions produce neutrons whilst others absorb them – otherwise the chain reaction will not be sustained.

Nuclear fissions generate considerable amounts of energy which is mostly turned into heat and so a nuclear reactor can be regarded as analogous to a boiler in a conventional power station.

There are a wide range of possible technologies for producing a nuclear reactor, of which only a handful have reached commercial or near-commercial status. These can be most conveniently categorised according to their moderator – light water, heavy water, graphite or no moderator. This book is primarily concerned with three types of reactor: two light water moderated reactors (LWRs) and one heavy water moderated reactor (HWRs) (see Table A.1).

The function of the moderator is to slow down the neutrons emitted in a fission which increases the probability that the neutron will fission another nucleus before it escapes from the core. The more efficient a moderator is, the higher the proportion of neutrons colliding with it which are not absorbed. Heavy water† and carbon (in graphite form) are the best moderators while ordinary or light water is less efficient but considerably cheaper. Fast reactors, which are not considered in detail here, have no moderator but surround the core with a 'blanket' of uranium which absorbs the neutrons and is itself turned into fissile material, mostly plutonium 239 (^{239}P) which can be recovered in a reprocessing plant and used as reactor fuel.

---

* The uranium 235 nucleus has 92 protons and 143 neutrons – the number 235 signifies the sum of the number of protons and neutrons. Other forms (isotopes) of uranium have different numbers of neutrons but the same number of protons, for example ^{238}U has 146 neutrons and 92 protons.

† Water is a compound of two atoms of hydrogen and one of oxygen. In heavy water, the two hydrogen atoms are replaced by a heavier isotope of hydrogen, deuterium.

Table A.1. *Technical characteristics of thermal reactors*

	PWR	BWR	CANDU
Moderator	Light water	Light water	Heavy water
Coolant	Pressurised light water	Boiling light water	Pressurised heavy water
Fuel	3% enriched uranium	3% enriched uranium	Natural uranium
Reactor vessel	Thick steel pressure vessel	Thinner steel vessel	Several hundred pressure tubes
Refuelling	Off-load	Off-load	On-load

The choice of moderator determines whether the proportion of the fissile isotope of uranium must be increased above its natural level of about 0.7% – in natural uranium 99.3% is the non-fissile ^{238}U.

In reactors which use light water as the moderator, this proportion must be increased to about 3% if a chain reaction is to be sustained whereas for heavy water reactors and some graphite moderated reactors natural uranium is used. Uranium enrichment is an expensive, technologically demanding process carried out in very few countries.

High-temperature reactors, which are not considered in detail here, use a different fuel system with very highly enriched uranium (93%) and thorium 232 (^{232}Th) which in the reactor is converted to fissile material.

The coolant removes the heat from the core directly or indirectly and is turned into steam which is used to drive the turbine generators which produce the electricity. The commercially available water moderated (light or heavy water) reactors use the moderating liquid as the coolant while graphite moderated reactors use gas (helium or carbon dioxide) as coolant. In the boiling water reactor (BWR) the light water coolant is allowed to boil and powers the turbine directly. The main advantages of this set-up are of simplicity and the fact that the vessel holding the core does not need to be very thick. The disadvantage of this arrangement is that the coolant inevitably picks up some radioactive contamination in the core which is transmitted to the turbine generator. In the pressurised water reactor (PWR) the coolant light water is highly pressurised to prevent it from boiling at temperatures well above its normal boiling point. This pressurised water is passed through a heat exchanger (steam generator) in which steam to power the turbines is produced. This arrangement requires a thick, strong, pressure vessel to contain the pressurised coolant liquid.

The most common form of HWR, the CANDU (CANadian Deuterium Uranium reactor) has much in common with the PWR, using pressurised heavy water as the coolant, but in the CANDU the fuel is contained in several hundred separate pressure tubes. These are much easier to manufacture than a single large pressure vessel and whereas the failure of a pressure vessel might be catastrophic, a failed pressure tube can be relatively easily withdrawn and replaced. Whereas both types of LWR must be shut down for refuelling (for three weeks or more) the CANDU can be refuelled whilst at full power. The pros and cons of on-load or off-load refuelling are complex but in general an on-load refuelled reactor should have a somewhat higher availability at the expense of the capital cost of the additional equipment necessary for on-load refuelling.

In terms of their commercial history, both types of LWR were primarily developed in the USA whilst the CANDU was developed in Canada. Table A.2 shows the sales patterns of the three principal current designs of reactor including their primary vendors, licensees and geographic distribution.

The factors lying behind patterns are one of the foci of this book but at this stage it is important to notice the dominance of the PWR. Whilst there have been substantial numbers of orders for BWRs, few have been placed recently and only Japan has a reactor ordering strategy involving significant numbers of BWR orders.

Table A.2. *Commercial development of the major reactor types*[a]

Technology	Vendor	Licensee (previous or current)	No. of units in operation (at 1.1.85)	No. of units on order (at 1.1.85)	Countries supplied (no. of units completed/ no. of units under construction)
Pressurised water reactor	Westinghouse (USA)		53	27	USA (34/20) Brazil (1/0) Italy (1/0) Japan (3/0) South Korea (2/4) Philippines (1/0) Spain (4/2) Sweden (3/0) Switzerland (2/0) Taiwan (1/1) Yugoslavia (1/0)
		Framatome[b] (France)	38	24	France (34/21) Belgium (3/0) South Korea (0/2) South Africa (1/1)
		Siemens/KWU[b] (FR Germany)	9	6	FR Germany (7/5) Netherlands (1/0) Spain (0/1) Switzerland (1/0)
		Acecowen (Belgium)	2	2	Belgium (2/2)
		Mitsubishi (Japan)	8	6	Japan (8/6)
	Babcock & Wilcox (USA)		0		USA (10/2)
		Babcock Brown Boveri (FR Germany)	0	1	FR Germany (0/1)
	Combustion Engineering (USA)				
	USSR		2	0	USA (11/4) Finland (2/0)
Total PWRs			112	66	

Type	Manufacturer			Countries
Boiling water reactor	General Electric (USA)	43	13	USA (28/9), FR Germany (1/0), India (2/0), Italy (1/0), Japan (3/0), Mexico (0/2), Spain (2/2), Switzerland (2/0), Taiwan (4/0)
	AEG/KWU (FR Germany)	9	0	FR Germany (8/0), Austria[c] (1/0)
	Ansaldo (Italy)	1	2	Italy (1/2)
	Hitachi (Japan)	3	2	Japan (3/2)
	Toshiba (Japan)	6	4	Japan (6/4)
	Asea Atom (Sweden)	9	2	Sweden (7/2), Finland (2/0)
Total BWRs		71	23	
Pressurised heavy water reactor	Atomic Energy of Canada	18	11	Canada (14/9), Argentina (1/0), India (2/0), South Korea (1/0)
	Dept. of Atomic Energy[d] (India)	1	5	India (1/5)
	Siemens/KWU	1	1	Argentina (1/1)
Total PHWRs		20	17	

[a] Includes all PWRs, BWRs and PHWRs of over 150 MW which have been completed or which are under construction in the world outside communist areas. Units which have been cancelled or on which work has been indefinitely suspended are excluded. Also excluded are all gas-cooled reactors.

[b] Framatome and Siemens/KWU terminated their licences with Westinghouse in 1982 and 1970 respectively.

[c] The unit built in Austria was completed but has never been operated due to a political decision.

[d] Atomic Energy of Canada terminated their agreement with the Department of Atomic Energy following the 'peaceful nuclear explosion' in India in 1974.

APPENDIX 2: DATA SOURCES

Plant performance analysis

The data required to analyse plant performance in the depth attempted in this book are as follows:

(a) plant name, unit number, country of installation and vendor
(b) date of first commercial operation
(c) net authorised plant output rating (MW) for the year in question
(d) net electricity generated during the year
(e) duration and cause of any plant shutdowns
(f) net design plant output rating (MW)

The source for items (a)–(e) is the annual publication *Operating experience with nuclear power stations in member states*, IAEA. This reports these data for all units in countries in the world outside communist areas (WOCA) which are members of the International Atomic Energy Agency (IAEA). In practice this includes all units in the WOCA except for those in Taiwan.

Item (f) is somewhat more problematic, there being considerable disagreement on the appropriate figure for a small number of problem units. For the majority of units there is agreement between *Operating experience with nuclear power stations in member states*, and other sources such as, *Nuclear Engineering International*'s 'Power reactor supplement' annual (1976 onwards). Where this is not the case the author has used his judgement to choose from this and numerous other sources the most appropriate value.

This IAEA data is available for the years 1973 to 1984 inclusive. For earlier years, the equivalent IAEA data is not available in the detail required although the gross electricity generated is published. From this data using the gross design plant output rating, it is possible to calculate the plant's capacity factor for those years. For 1985 to 1986, *Nucleonics Week* publishes gross electricity generation figures which can be treated similarly. Gross design ratings are derived from the same sources and using the same methods as net design ratings.

Projected plant completion dates

Due to changes in construction schedules, expected plant completion dates are continually changing (usually slipping) and so no one source is adequate for this data. The data used in this book are the most recently published dates from a reliable (preferably utility) source in journals such as *Nucleonics Week*, and *Nuclear Engineering International*.

Other plant data

Other plant data such as date of order, date of start of construction etc. is not contained reliably in any one source. For such items, the sources are indicated in the text.

References

Arrow, K. J. (1962). The economic implications of learning by doing. *Review of Economic Studies*, **XXIX**, 2 (June 1962), pp. 155–73.

Bacher, P. (1986). How EdF cuts the time to full power. *Nuclear Engineering International*, **31**. No. 384, pp. 41–2.

Baumier, J. & Bergougnoux, J. (1982). *The costs of nuclear energy in France*, IAEA Conference on Nuclear Power Experience, Vienna, September 1982, p. 2.

Bellet, J., Houyez, A., Journet, J. & Pierrard, J. H. (1985). The N4 plant: culmination of French PWR experience. *Nuclear Engineering International*, **30**, No. 365, pp. 26–8.

Bennett, L. L., Karousakis, P. M. & Moynet, G. (1982). *Review of nuclear power costs around the world*, Paper presented to the IAEA Conference on Nuclear Power Experience, Vienna, September 1982, p. 2.

Buckley, C. M., MacKerron, G. S. & Surrey, A. J. (1980). The international uranium market. *Energy Policy*, **8**, No. 2, pp. 84–104.

Bupp, I. C. & Derian, J. C. (1981). *The failed promise of nuclear power: the story of light water*. Basic Books, New York.

Central Electricity Generating Board (1982). *CEGB statement of case*, CEGB, London, Appendix L, Table L7, p. 12.

Crowley, J. H. & Kaminski, R. S. (1985). What the USA can learn from France about cost control. *Nuclear Engineering International*, **30**, No. 371, pp. 34–6.

Dawson, F. G. (1976). *Nuclear power: development and management of a technology*. University of Washington Press, Seattle.

Electric Power Research Institute (1984). *An analysis of power plant construction lead times*, 2 vols, EPRI E–2880, Palo Alto.

Energy Information Administration (annual). *Historical plant cost and annual production expenses for selected electric plants*. US Department of Energy, Washington DC.

Energy Information Administration (1982). *Federal support for nuclear power reactor design and fuel cycle*. US Department of Energy, EIA–0201/13, Washington DC.

Financial Times (1985). Electricity leads VEBA profit. European Energy Report No. 193, p. 15.

Fouquet, D. (1985). *The development of competitive uses of electricity in France*. Paper presented to the Benelux Association of Energy Economists, Luxembourg, September 23–25, 1985.

Hansen, U. (1984). *Nuclear economics in the Federal Republic of Germany.* Paper presented to the 6th International Conference of the International Association of Energy Economists, Cambridge, 9–11 April, 1984.

Hellman, R. & Hellman, C. J. C. (1982). *The competitive economics of nuclear and coal power*, Lexington Books, Lexington, Mass., p. 157.

International Atomic Energy Agency (annual). *Nuclear power reactors in the world.* International Atomic Energy Agency, Vienna.

International Energy Agency (1985a). *Energy balances of OECD countries – 1982–83.* OECD, Paris.

International Energy Agency (1985b). *Energy research, development and demonstration in the IEA countries.* OECD, Paris, Table 5A, p. 39.

International Energy Agency (1986a). *Coal information 1986.* OECD, Paris.

International Energy Agency (1986b). *Natural gas prospects.* OECD, Paris.

Jestin-Fleury, H. (1985). *The principal determinants of competition between gas and electricity in the French economy.* Paper presented to the Benelux Association of Energy Economists, Luxembourg, 23–25 September, 1985.

Joskow, P. L. & Rozanski, G. A. (1979). The effects of learning by doing on nuclear plant operating reliability. *The Review of Economics and Statistics*, **LXI**, No. 2, pp. 161–8.

Keck, O. (1981). *Policymaking in a nuclear program: the case of the West German fast breeder reactor.* Lexington Books, Lexington, Mass.

Kim, C. H. & Chung, C. H. (1979). Cost comparison of PWR and PHWR nuclear power plants in Korea. *Journal of the Korean Nuclear Society*, **11**, No. 4, pp. 263–74.

Komanoff, C. (1976). *Power plant performance: nuclear and coal capacity factors and economics.* Council on Economic Priorities, New York.

Komanoff, C. (1981). *Power plant cost escalation.* Van Nostrand Reinhold, New York.

Lester, R. K. (1982). Nuclear power plant lead times. In. I. Smart (ed.), *World nuclear energy.* Johns Hopkins, Baltimore, pp. 123–47.

Lortie, P. & Schweitzer, R. (1981). *A strategy for the development and strengthening of the Canadian nuclear industry.* SECOR Inc., Montreal.

Lucas, N. J. D. (1979). *Energy in France: planning, politics and policy.* Europa Publications, London.

Macdonald, D. S. (1974). Canada's role in world developments of nuclear energy. *Nuclear Engineering International*, **19**, No. 217, p. 475.

Marquis, G. (1982). *Experience with nuclear power plant investment costs in the Federal Republic of Germany and expected future electricity production costs.* Paper presented to the International Conference on Nuclear Power Experience, IAEA–CIN–42/393, Vienna, 13–17 September, 1982.

Masters, R. (1979). Progress report on France. *Nuclear Engineering International*, **24**, No. 282, pp. 47–62.

Mooz, W. (1981). Cost analysis for LWR power plants. *Energy*, **6**, No. 3, pp. 197–225.

Mullenbach, P. (1963). *Civilian nuclear power: economic issues and policy formation.* The Twentieth Century Fund, New York.

Nelkin, D. & Pollak, M. (1981). *The atom besieged: extraparliamentary dissent in France and Germany.* MIT Press, Cambridge, Mass.

Nuclear Energy Agency (1980). *Description of licensing systems and inspection of nuclear installations*. OECD, Paris.

Nuclear Energy Agency (1983). *The cost of generating electricity in nuclear and coal fired power stations*. OECD, Paris.

Nuclear Engineering International (1982a). COGEMA makes a loss while devaluation hits EdF. **27**, No. 331, p. 4.

Nuclear Engineering International (1982b). EdF replaces control rod guide tubes. **27**, No. 334, p. 2.

Nucleonics Week (1980). The time has come for Canada to follow France's lead and go flat out on nuclear. **21**, No. 39, pp. 7–8.

Nucleonics Week (1983). EdF changing bad pins without scrapping whole control rod guide tube. **24**, No. 2, pp. 7–8.

Nucleonics Week (1984a). French demonstrate load following for international audience. **25**, No. 44, pp. 3–4.

Nucleonics Week (1984b). Key to cheaper French nuclear plants is labor use, study shows. **25**, No. 48, pp. 9–10.

Nucleonics Week (1985a). KWU takes exception to Framatome claims of lead in load following. **26**, No. 3, pp. 9–10.

Nucleonics Week (1985b). RWE rates mount as nuclear ambitions are fettered by coal lobby. **26**, No. 33, pp. 8–10.

Nucleonics Week (1986a). EdF will have excess nuclear units in 1990, but chairman doesn't mind. **27**, No. 4, p. 7.

Nucleonics Week (1986b). EdF finds new 'microcracks' in Dampierre–1's steam generator. **27**, No. 8, pp. 2–3.

Nucleonics Week (1986c). Economic woes threaten Laguna Verde as unit 1 completion nears. **27**, No. 41, pp. 11–13.

Ontario Royal Commission on Electric Power Planning (1980). *Report*. Royal Commission on Electric Power Planning, Toronto.

Patterson, W. C. (1984). *The plutonium business and the spread of the bomb*. Paladin Books, London.

Perry, R. with Alexander, A. J., de Leon, P., Gandara, A., Mooz, W. E., Rolph, E., Siegel, S. & Solomon, K. A. (1977). *Development and commercialisation of the light water reactor, 1946–1976*. R–2180–NSF, Rand Corporation, Santa Monica.

President's Commission (1979). *The legacy of TMI*, Report of the President's Commission on the accident at Three Mile Island. Pergamon Press, New York.

Roth, E. B. (1982). *The development and commercialisation of light water reactors in the United States*. ECON Inc, Princeton.

Schmidt, R. (1984). *Assessing prospects for smaller reactors*. IAEA Bulletin, **26**, No. 4, pp. 29–35.

Select Committee on Energy (1981). *First Report Session 1980–81: The Government's statement on the new nuclear power programme*. HC114–ii, HMSO, London, p. 453.

Sort, M. (1985). France prepares for steam generator replacement. *Nuclear Engineering International*, **30**, No. 376, pp. 45–7.

Surrey, A. J. & Chesshire, J. H. (1972). *The world market for electric power equipment*. Science Policy Research Unit, University of Sussex, Brighton.

Surrey, A. J. & Thomas, S. D. (1980). Worldwide nuclear plant performance: lessons for technology policy. *Futures*, **12**, No. 1, pp. 3–17.

Syndicat CFDT de l'Energie Atomique (1980). *Le dossier electronucleaire*. Sciences, Paris.

Tanguy, P. (1985). Safety and nuclear power plant standardisation: the French experience. *Public Utilities Fortnightly*, **16**, No. 9, pp. 20–5

Torresi, F. (1972). Recent structural reorganisation within the French nuclear industry. *Nuclear Engineering International*, **17**, No. 196, pp. 702–3.

United States Atomic Energy Commission (1968). *Current status and future technical and economic potential of light water reactors*. WASH–1082, USAEC, Washington. This series continued through WASH–1150 (1970), WASH–1230 (1972), and WASH–1345 (1974).

United States Atomic Energy Commission (1974). *Power plant capital costs: current trends and sensitivity to economic parameters*. WASH–1345, USAEC, Washington, pp. 27–30.

United States Atomic Energy Commission (1975). *Reactor safety study: an assessment of accident risks in US commercial power plants*. WASH–1400, USAEC, Washington.

United States National Technical Information Service (undated). Phase IV update (1983): US national report for the energy economic data base program, US Dept. of Commerce, Washington.

Walker, W. B. & Lonnroth, M. (1983). *Nuclear power struggles: industry competition and proliferation control*. George Allen & Unwin, London.

Wilkie, T. (1980). Cracks in French pressure vessels pose no danger. *Nuclear Engineering International*, **25**, No. 294, pp. 27–9.

Williams, R. (1980). *The nuclear power decisions: British policies 1953–78*. Croom Helm, London.

Wilson, A. (1982). *CEGB proof of evidence on: construction time, cost and operating performance of PWR, AGR and coal-fired generating plant*. CEGB, London.

Winnacker, K. & Wirtz, K. (1975). *Nuclear energy in Germany*. American Nuclear Society, Illinois.

Yu, A. M. & Bate, D. L. S. (1982). *Trends in the capital costs of CANDU generating stations*. Paper presented to the IAEA Conference on nuclear power experience, Vienna, 13–17 September, 1982 (IAEA–CN–42/141).

Zimmerman, M. (1981). *Learning effects of the commercialisation of new energy technologies: the case of nuclear power*. Working paper, Sloan School of Management, MIT, Mass.

Index